SCHOLASTIC

GCSE 9-1

FOUNDATION

MATHEMATICS

REVISION GUIDE

FOR ALL EXAM BOARDS

Gwenllian Burns and
Catherine Murphy

Authors Gwenllian Burns and Catherine Murphy

Editorial team Haremi Ltd

Series designers emc design ltd

Typesetting York Publishing Solutions Pvt. Ltd.

Illustrations York Publishing Solutions Pvt. Ltd.

App development Hannah Barnett, Phil Crothers and Haremi Ltd

Designed using Adobe InDesign

Published by Scholastic Education, an imprint of Scholastic Ltd, Book End, Range Road, Witney, Oxfordshire, OX29 0YD

Registered office: Westfield Road, Southam, Warwickshire CV47 0RA

www.scholastic.co.uk

Printed by Bell & Bain Ltd, Glasgow

© 2017 Scholastic Ltd

1 2 3 4 5 6 7 8 9   7 8 9 0 1 2 3 4 5 6

British Library Cataloguing-in-Publication Data

A catalogue record for this book is available from the British Library.

ISBN 978-1407-16909-5

Acknowledgements

Map on page 103: © Petersfield Town Council (contains Ordnance Survey data © Crown copyright and database right 2010)

Every effort has been made to trace copyright holders for the works reproduced in this book, and the publishers apologise for any inadvertent omissions.

Note from the publisher

Please use this product in conjunction with the official specification and sample assessment materials for the exam board that will be setting your examinations. Ask your teacher if you are unsure where to find them.

Contents

Topic 4

GEOMETRY AND MEASURES

Topic 5

PROBABILITY

Topic 6

STATISTICS

How to use this book

This Revision Guide has been produced to help you revise for your 9–1 GCSE in Foundation Mathematics. Broken down into topics and subtopics it presents the information in a manageable format. Written by subject experts to match the new specifications, it revises all the content you need to know before you sit your exams.

The best way to retain information is to take an active approach to revision. Don't just read the information you need to remember – do something with it! Transforming information from one form into another and applying your knowledge through lots of practice will ensure that it really sinks in. Throughout this book you'll find lots of features that will make your revision an active, successful process.

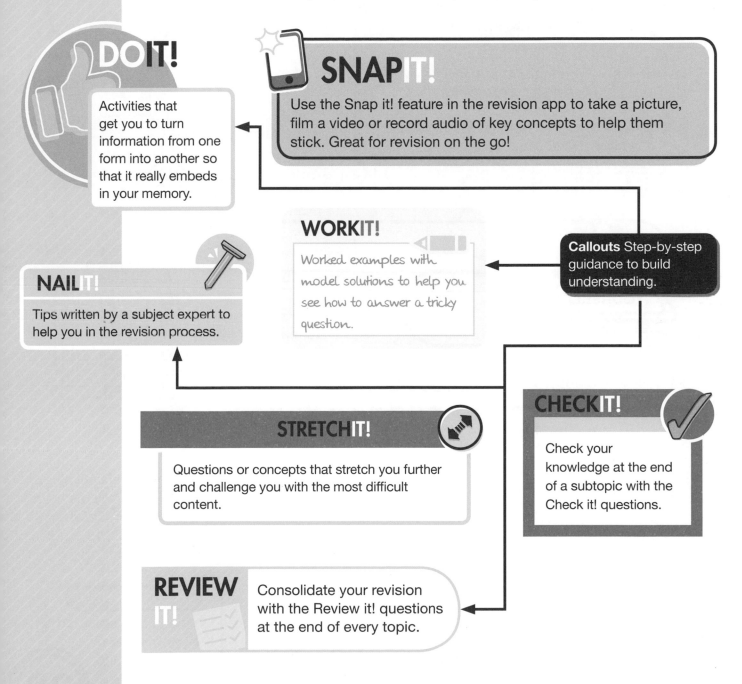

DO IT!

Activities that get you to turn information from one form into another so that it really embeds in your memory.

SNAP IT!

Use the Snap it! feature in the revision app to take a picture, film a video or record audio of key concepts to help them stick. Great for revision on the go!

WORK IT!

Worked examples with model solutions to help you see how to answer a tricky question.

Callouts Step-by-step guidance to build understanding.

NAIL IT!

Tips written by a subject expert to help you in the revision process.

STRETCH IT!

Questions or concepts that stretch you further and challenge you with the most difficult content.

CHECK IT!

Check your knowledge at the end of a subtopic with the Check it! questions.

REVIEW IT!

Consolidate your revision with the Review it! questions at the end of every topic.

Use the All Boards Foundation Mathematics Exam Practice Book alongside the Revision Guide for a complete revision and practice solution. Packed full of exam-style questions for each subtopic, along with complete practice papers, the Exam Practice Book will get you exam ready!

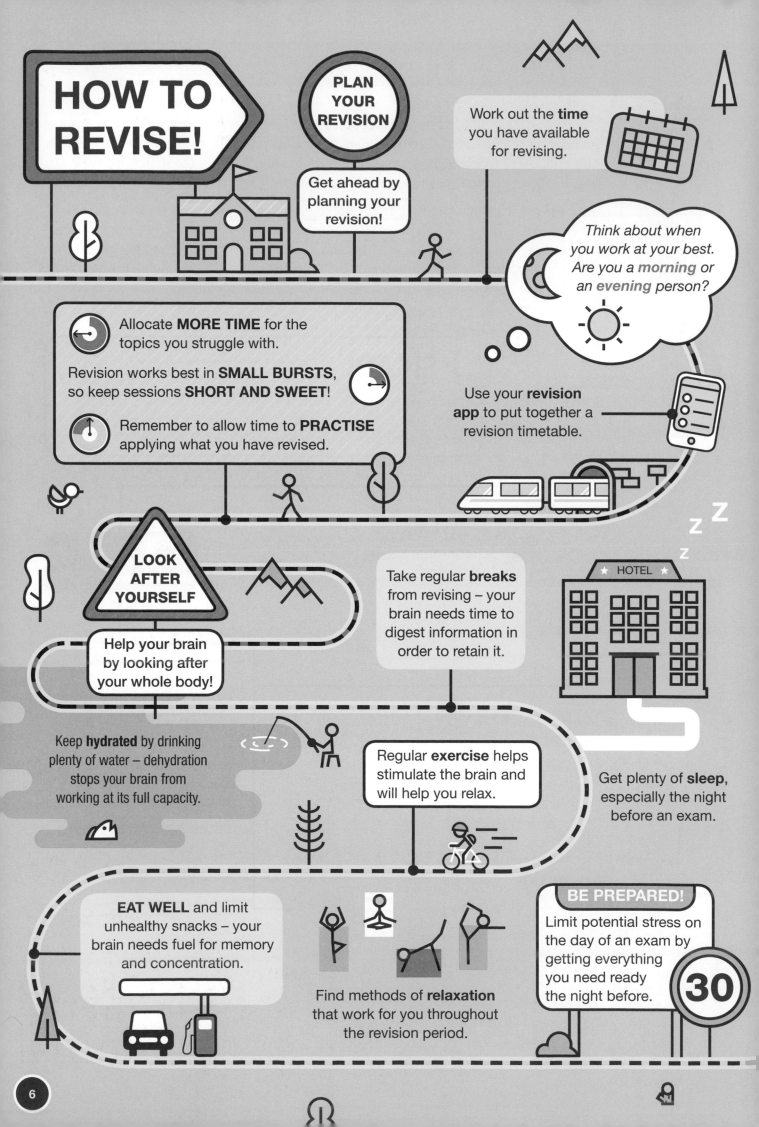

HOW TO REVISE!

PLAN YOUR REVISION

Get ahead by planning your revision!

Work out the **time** you have available for revising.

*Think about when you work at your best. Are you a **morning** or an **evening** person?*

Allocate **MORE TIME** for the topics you struggle with.

Revision works best in **SMALL BURSTS**, so keep sessions **SHORT AND SWEET**!

Remember to allow time to **PRACTISE** applying what you have revised.

Use your **revision app** to put together a revision timetable.

LOOK AFTER YOURSELF

Help your brain by looking after your whole body!

Take regular **breaks** from revising – your brain needs time to digest information in order to retain it.

HOTEL

Keep **hydrated** by drinking plenty of water – dehydration stops your brain from working at its full capacity.

Regular **exercise** helps stimulate the brain and will help you relax.

Get plenty of **sleep**, especially the night before an exam.

EAT WELL and limit unhealthy snacks – your brain needs fuel for memory and concentration.

Find methods of **relaxation** that work for you throughout the revision period.

BE PREPARED!

Limit potential stress on the day of an exam by getting everything you need ready the night before.

30

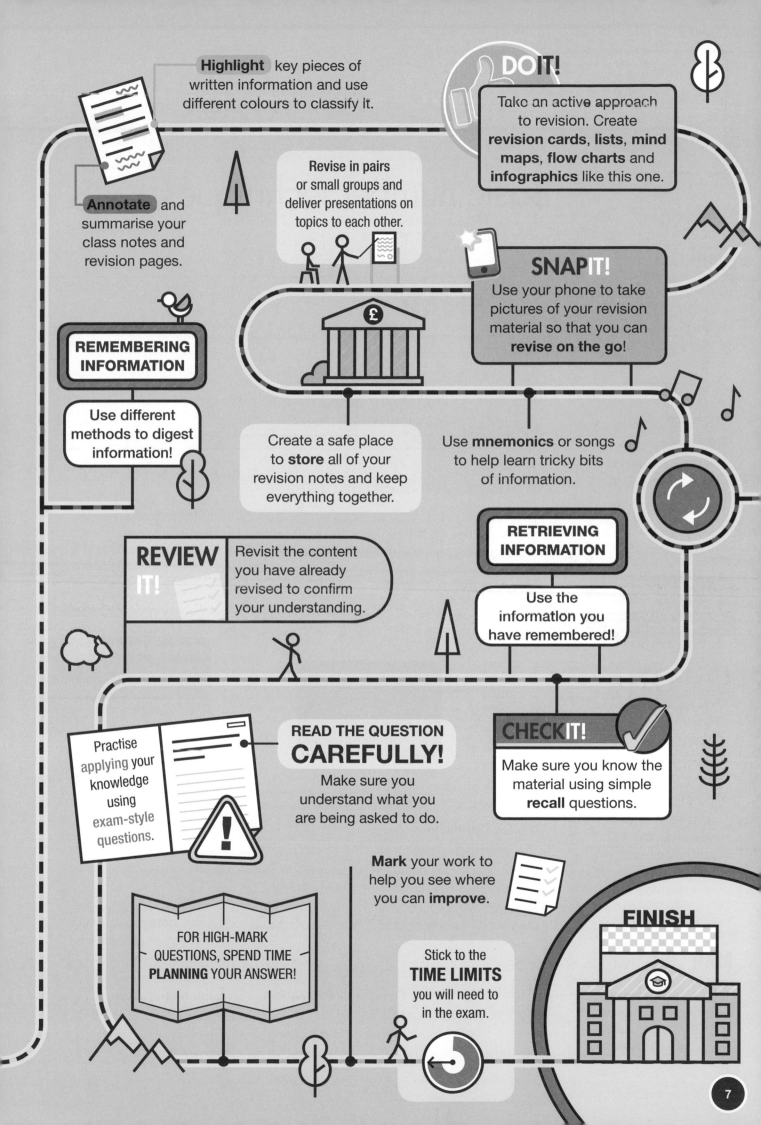

Number

Basic number techniques

SNAP IT!

Even numbers can be divided evenly into groups of two: they are all multiples of 2. The even numbers are 2, 4, 6, 8, 10, …

Odd numbers cannot be divided evenly into groups of two. The odd numbers are 1, 3, 5, 7, 9, …

NAIL IT!

Addition

You may be asked to:
- find the sum
- work out the total
- add together.

These all mean the same thing – you have to add the values!

Subtraction

You may be asked to:
- find the difference.

This means subtract the values.

Addition and Subtraction

To add or subtract numbers you can use the column method. Remember to correctly line up the digits of the number so that the units (ones), tens, hundreds, etc. are directly beneath one another.

WORKIT!

Find the value of 2940 − 193 + 28311

First work out: 2940 − 193

You cannot subtract 9 from 3 so take from the hundreds column.

$$
\begin{array}{r}
^{8}\,^{13}\!\!^{1}\\
2\,9\,4\,0\\
-\ \ 1\,9\,3\\
\hline
2\,7\,4\,7
\end{array}
$$

You cannot subtract 3 from 0 so take '1' from the tens column, which becomes '10' in the units column.

Then add 28311.

$$
\begin{array}{r}
2\,7\,4\,7\\
+\ 2\,8\,3\,1\,1\\
\hline
3\,1\,0\,5\,8\\
^{1}\ ^{1}
\end{array}
$$

Make sure you line the addition up correctly – it doesn't matter with addition which number goes on top of the calculation.

DOIT!

There are 23 456 people at a concert one night and only 1894 the next night.

a How many attended over the two nights?

b What was the difference in the number who attended on each night?

NAIL IT!

Inverse Operations

The inverse of addition is subtraction and the inverse of multiplication is division.

To check you have carried out a calculation correctly you could 'undo' the operation by carrying out the inverse operation.

For example if you have worked out that 345 − 99 = 246 you could check your calculation by working out 99 + 246. If you don't get 345 something has gone wrong.

Ordering positive and negative integers

A negative number is smaller than zero.

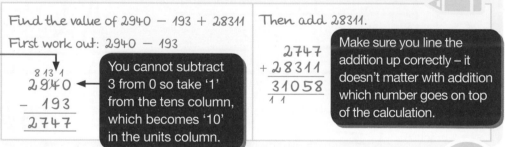

As you move left along the number line, the numbers get lower.

−10 is lower than −4.

NAIL IT!

If the temperature is −10°C it will be colder than a temperature of −1°C, so −10 is lower than −1.

Place value

In a number, for example 3 108 475.926, the value of each digit depends on its position in the number − its **place value**.

millions	hundreds of thousands	tens of thousands	thousands	hundreds	tens	units	tenths	hundredths	thousandths
3	1	0	8	4	7	5 • 9		2	6

The 3 is worth 3 million: 3 000 000

The 7 is worth 7 tens: 70

The 2 is worth 2 hundredths: 0.02

> The units are the numbers from 1 to 9. You may also know them as the 'ones'.

DOIT!

Write down a large number using all the digits 1−9. Write down the value of each digit.

Ordering decimals

To write decimals in order of size, compare the whole number part, then tenths, then hundredths and so on. Work from left to right.

WORKIT!

a Which is larger, 3.431 or 3.9?

   3.9 is larger than 3.431 ◄

b Put these numbers in ascending order.

   −3, 3.03, 0.3, −0.03

   −3, −0.03, 0.3, 3.03 ◄

> First look at the whole number parts:
> **3**.431 or **3**.9   they are the same.
> Next compare the tenths:
> 3.**4**31 or 3.**9**   9 is greater than 4.

> Think about the position of each number on the number line.

NAILIT!

Ascending means going up.

Descending means going down.

SNAPIT! Comparing numbers

To compare numbers, use the symbols:

Symbol	Meaning	Symbol	Meaning
=	equal to	≠	not equal to
<	less than	>	greater than
≤	less than or equal to	≥	greater than or equal to

NAILIT!

The wide end (or 'mouth') of the inequality sign always points towards the larger value.

For example, 15 > 2 means '15 is greater than 2'.

CHECKIT!

> You can also use a wavy equals sign: ≈ to mean 'is approximately equal to'.

1 Decide whether the following statements are true or false.

   a −4 < −7   d 3.41 ≥ 3.14

   b 4.1 > 4.01   e 0.909 ≤ 0.99

   c 0.1 ≠ 0.01

2 Write these numbers in **descending** order:

   −2.5, −4.2, −0.3, −1.5, −7.2

3 Write these numbers in **ascending** order:

   0.412, 0.124, 0.442, 0.049, 1.002

4 Replace each ? symbol with > or <.

   a 0.25 ? 0.52

   b 1.07 ? 1.1

   c 13.4 ? 13.099

Factors, multiples and primes

NAIL IT!

To find all the factors of a number, list all the pairs that multiply to give that number. Start with 1 and work up!

For example, factors of 20:

1 × 20

2 × 10

4 × 5

SNAP IT! Factors, multiples and primes

Factors

Factors of a number divide into it exactly. For example, the factors of 12 are 1, 2, 3, 4, 6 and 12. Notice that you include 1 and the number itself.

Multiples

To find the multiples of a number, multiply it by 1, 2, 3 and so on. For example, multiples of 5 are 5, 10, 15, 20, 25 and so on.

Prime numbers

A prime number is one with **exactly** two factors. For example, 7 is prime as its only factors are 1 and 7.

1 is not prime, because it only has 1 factor (itself). So the lowest prime number (and the only even prime) is 2

WORKIT!

Find the highest common factor of 20 and 36.

The factors of 20 are: 1, 2, ④, 5, 10, 20

The factors of 36 are: 1, 2, 3, ④, 6, 9, 12, 18, 36

The highest factor that occurs in both lists is 4.

The highest common factor of 20 and 36 is 4.

Highest common factor

The highest common factor (HCF) of two numbers is the largest number that is a factor of both (divides into both exactly).

Lowest common multiple

The lowest common multiple (LCM) of two numbers is the smallest number that is a multiple of both (they will both divide into it exactly).

WORKIT!

Find the lowest common multiple of 10 and 15.

Multiples of 10 are: 10, 20, ㉚, 40, 50, 60, 70, 80...

Multiples of 15 are: 15, ㉚, 45, 60, 75...

The lowest value that appears in both lists is 30.

So the lowest common multiple is 30.

NAIL IT!

To find the lowest common multiple, list as many multiples as necessary until you find one that appears in both lists. Sometimes it is simply the result of multiplying the two numbers together.

For example, the lowest common multiple of 5 and 7 is 5 × 7 = 35.

Prime factor decomposition

All numbers can be written as a product of their prime factors. This is called prime factor decomposition.

WORKIT!

Write 180 as a product of prime factors. Write your answer in index form (using indices).

Here is the factor tree for 180.

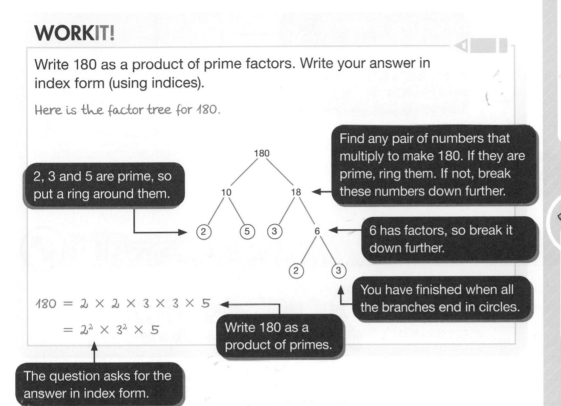

2, 3 and 5 are prime, so put a ring around them.

Find any pair of numbers that multiply to make 180. If they are prime, ring them. If not, break these numbers down further.

6 has factors, so break it down further.

You have finished when all the branches end in circles.

$180 = 2 \times 2 \times 3 \times 3 \times 5$
$= 2^2 \times 3^2 \times 5$

Write 180 as a product of primes.

The question asks for the answer in index form.

DOIT!

Choose two different two-digit numbers between 10 and 30. Find the highest common factor and the lowest common multiple of the two numbers.

 NAILIT!

Never include a 1 in the factor tree, since 1 is not a prime number.

You can use the prime factor decomposition to find the highest common factor and the lowest common multiple of two numbers.

To find the highest common factor and lowest common multiple of 30 and 40, first write both numbers as a product of primes, using prime factor decomposition.

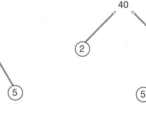
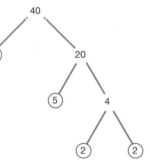

Display the Information in a Venn diagram, making sure that any numbers that appear in both trees are written in the intersection (the leaf-shaped area where the circles overlap).

The highest common factor is found by multiplying together any factors that appear in **both** trees. These are the ones in the intersection of the Venn diagram.

$HCF = 2 \times 5 = 10$

The lowest common multiple is found by multiplying all the factors in the Venn diagram:

$LCM = 3 \times 2 \times 5 \times 2 \times 2 = 120$

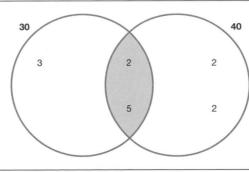

See page 187 for more explanation of set notation.

WORKIT!

Given the following facts, find the highest common factor and lowest common multiple of 18, 20 and 30.

$18 = 2 \times 3 \times 3$

$20 = 2 \times 2 \times 5$

$30 = 2 \times 3 \times 5$

Draw a Venn diagram to display the information.

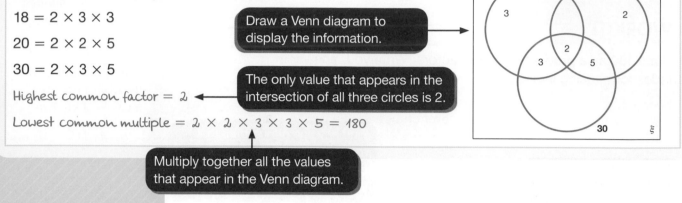

Highest common factor = 2

The only value that appears in the intersection of all three circles is 2.

Lowest common multiple = $2 \times 2 \times 3 \times 3 \times 5 = 180$

Multiply together all the values that appear in the Venn diagram.

STRETCHIT!

Prime factors can be small but they may also be very large. Although you won't be asked to find large prime factors in an exam, they are very useful for encrypting and decoding digital data. When a number only has huge prime factors they are extremely difficult to find!

CHECKIT!

1 Here is a list of five numbers:

1   5   12   30   45

Which numbers in the list are

a prime

b factors of 12

c **not** multiples of 2

d divisors of 10?

2 Find the highest common factor and lowest common multiple of 150 and 70.

3 Write 90 as a product of prime factors using index form.

Index form means numbers with powers.

The product is what you get when you multiply things together.

4 Given that

$120 = 2 \times 2 \times 2 \times 3 \times 5$

$70 = 2 \times 5 \times 7$

$30 = 2 \times 3 \times 5$

a find the highest common factor

b find the lowest common multiple.

5 The highest common factor of a pair of numbers is 6. Both numbers are between 10 and 20. What are they?

Calculating with negative numbers

Finding differences

A number line is useful to find the difference between a positive and negative number.

The difference between −2 and 4 can be calculated by considering the 'jumps' that are taken along the number line to move from −2 to 4.

The difference between −2 and 4 is 2 + 4 = 6.

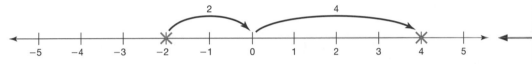

First jump from −2 to 0, then jump from 0 to 4.

Adding and subtracting negative numbers

SNAP IT!   Adding and subtracting negative numbers

Adding a negative number is the same as subtracting a positive number.
Subtracting a negative number is the same as adding a positive number.

WORKIT!

Work out:

a −4 + −2

  −4 − 2 = −6

Adding a negative number is equivalent to subtracting, so you can rewrite the calculation as a subtraction.

b 3 − −5

  3 + 5 = 8

Subtracting a negative number is equivalent to adding, so you can rewrite the calculation as an addition.

Multiplying and dividing negative numbers

SNAP IT!   Multiplying and dividing negative numbers

When multiplying and dividing two negative numbers, remember the rules:

- If the signs are the same the answer will be positive.
- If the signs are different the answer will be negative.

STRETCHIT!

If you multiply together three negative numbers, will the solution be positive or negative? Is this always true?

WORKIT!

Work out

a -2×8

$-2 \times 8 = -16$

> Multiply: $2 \times 8 = 16$
>
> The signs are different so the answer will be negative.

b $-20 \div -5$

$-20 \div -5 = 4$

> Divide: $20 \div 5 = 4$
>
> The signs are the same, so the answer is positive.

c $-3 \times -5 \times -1$

$-3 \times -5 \times -1 = 15 \times -1$

$= -15$

> Work out -3×-5 first.

✓ CHECKIT!

DOIT!

Make a mind map about negative numbers. Include everything you know about working with negative numbers.

1 Work out:

a $-8 + -3$ **d** $14 - -4$

b -9×-11 **e** $-8 + 8$

c $18 \div -3$ **f** $12 - -15 - 2$

2 Two of the numbers below are multiplied together.

Which pair of numbers gives the lowest answer?

| -5 | -8 | 4 | 9 |

3 The average daily temperature in Moscow in December is −8°C. At the North Pole the average daily temperature is −40°C.

How much colder is it at the North Pole than in Moscow?

Division and multiplication

Division

You can use short division to divide a large number by a single-digit number. The number you are dividing by is called the **divisor**. You work from left to right, dividing into each digit in turn, as shown in the Work it! example.

WORKIT!

Work out 348 ÷ 3.

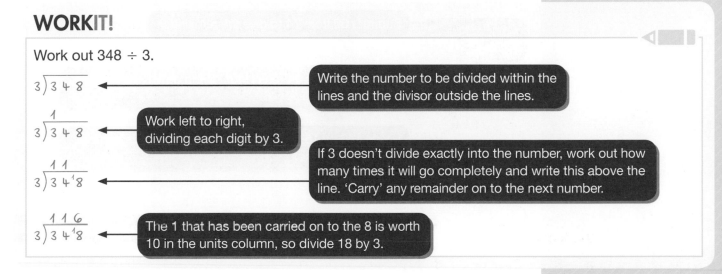

3$\overline{)348}$ ← Write the number to be divided within the lines and the divisor outside the lines.

$\overset{1}{3\overline{)348}}$ ← Work left to right, dividing each digit by 3.

$\overset{1\ 1}{3\overline{)3\ 4^{1}8}}$ ← If 3 doesn't divide exactly into the number, work out how many times it will go completely and write this above the line. 'Carry' any remainder on to the next number.

$\overset{1\ 1\ 6}{3\overline{)3\ 4^{1}8}}$ ← The 1 that has been carried on to the 8 is worth 10 in the units column, so divide 18 by 3.

Sometimes the divisor will not divide exactly. There are two ways to find the exact solution.

Method 1: Write the solution as a mixed number

$\overset{0\ 2\ 4}{5\overline{)1\ ^{1}2\ ^{2}4}}$ remainder 4

The remainder still needs to be divided by 5, so write:

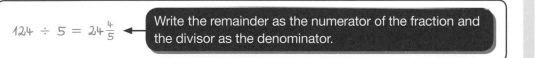

$124 \div 5 = 24\frac{4}{5}$ ← Write the remainder as the numerator of the fraction and the divisor as the denominator.

Method 2: Work out the solution as a decimal

$\overset{0\ 2\ 4\ .8}{5\overline{)1\ ^{1}2\ ^{2}4\ ^{4}0}}$ ← Insert a decimal point and write as many zeros as you like after the number. Keep working as before. When the divisor divides exactly you can stop adding more zeros.

$124 \div 5 = 24.8$

Long division

You can use long division to divide a large number by a two-digit or larger number. As with short division, you work from left to right. This time you take two or more digits at once (because your divisor is bigger than 9). See the Work it! on the next page for how to do this step by step.

DOIT!

To check the answer to a calculation you can use the inverse operation (see page 8).

To check that 348 ÷ 3 = 116, which calculation should you do?

116 × 3 or 116 ÷ 3?

NAILIT!

124 = 124.0
= 124.00 = 124.000

WORKIT!

Work out 746 ÷ 15.

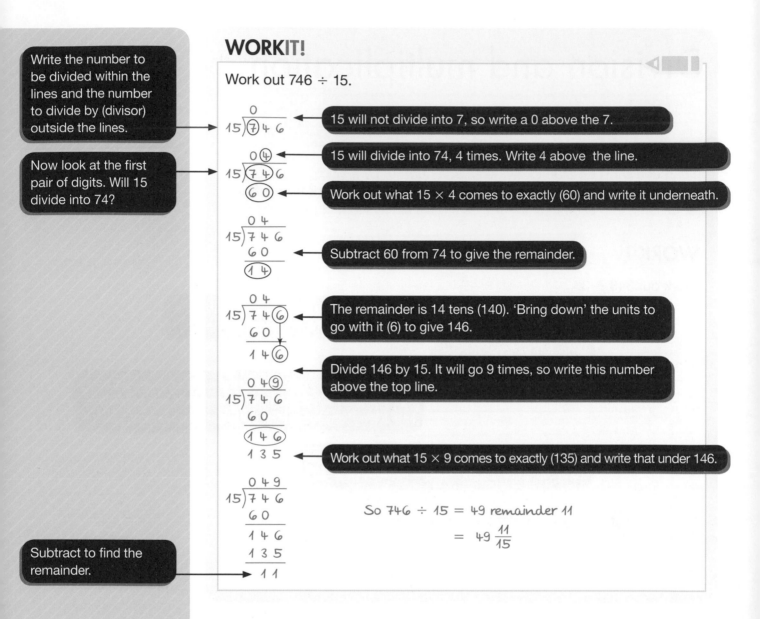

Write the number to be divided within the lines and the number to divide by (divisor) outside the lines.

15 will not divide into 7, so write a 0 above the 7.

Now look at the first pair of digits. Will 15 divide into 74?

15 will divide into 74, 4 times. Write 4 above the line.

Work out what 15 × 4 comes to exactly (60) and write it underneath.

Subtract 60 from 74 to give the remainder.

The remainder is 14 tens (140). 'Bring down' the units to go with it (6) to give 146.

Divide 146 by 15. It will go 9 times, so write this number above the top line.

Work out what 15 × 9 comes to exactly (135) and write that under 146.

So 746 ÷ 15 = 49 remainder 11

$$= 49\frac{11}{15}$$

Subtract to find the remainder.

Multiplication

Use short multiplication to multiply a one-digit number by a number with two or more digits.

WORKIT!

Work out 172 × 8.

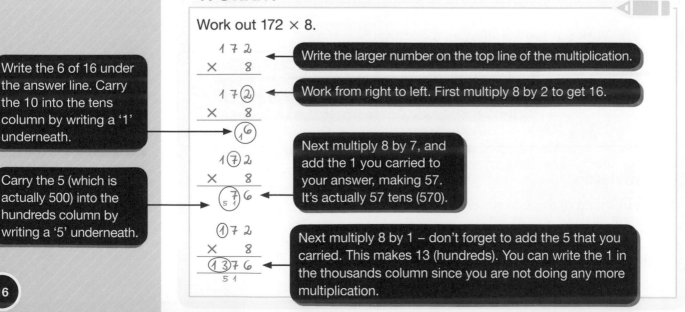

Write the larger number on the top line of the multiplication.

Work from right to left. First multiply 8 by 2 to get 16.

Write the 6 of 16 under the answer line. Carry the 10 into the tens column by writing a '1' underneath.

Next multiply 8 by 7, and add the 1 you carried to your answer, making 57. It's actually 57 tens (570).

Carry the 5 (which is actually 500) into the hundreds column by writing a '5' underneath.

Next multiply 8 by 1 – don't forget to add the 5 that you carried. This makes 13 (hundreds). You can write the 1 in the thousands column since you are not doing any more multiplication.

Long multiplication

Use long multiplication to multiply two numbers with more than one digit each.

WORKIT!

Work out 126 × 43.

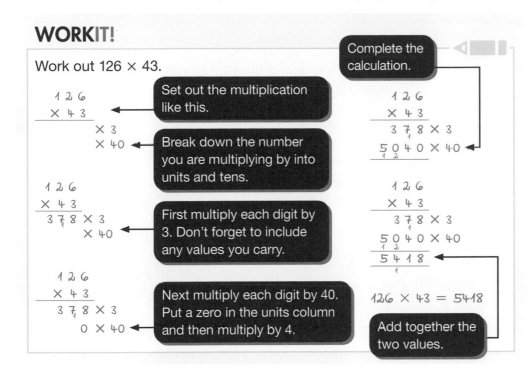

Set out the multiplication like this.

Break down the number you are multiplying by into units and tens.

First multiply each digit by 3. Don't forget to include any values you carry.

Next multiply each digit by 40. Put a zero in the units column and then multiply by 4.

Complete the calculation.

Add together the two values.

126 × 43 = 5418

DO IT!

Roll a dice three times to produce a 3-digit number. Roll it twice more to give a 2-digit number.

Divide the 3-digit number by the 2-digit number and check your answer on a calculator.

STRETCH IT!

To multiply by 239 you would need three rows: × 9
          × 30
          × 200

Work out 621 × 239.

For the first two questions, use inverse operations to check your calculations.

CHECKIT!

Do not use a calculator for these questions.

1 Work out:
  a 235 × 9   **b** 924 × 61

2 Work out:
  a 235 ÷ 5   **c** 2146 ÷ 17
  b 4128 ÷ 8

3 In a factory, 265 pencils are packed into boxes of 8.
  a How many boxes are filled?
  b How many pencils are left over?

4 Work out £365 ÷ 4.

5 A cinema charges £9 per ticket. If a class of 32 students goes to the cinema, what is the total cost of tickets?

6 Calculate 923 ÷ 3 and give your answer as a mixed number.

7 Work out 823 × 35.

8 Find the missing number in these calculations:
  a 239 + ☐ = 921
  b ☐ × 87 = 1131
  c 23 × ☐ + 8 = 123

9 Eggs are packed into boxes of 12. How many complete boxes can be filled with 450 eggs?

10 Aaron is calculating 726 × 52. Explain what he has done wrong.

Calculating with decimals

To add or subtract decimals, line up the decimal points.

WORKIT!

Aisha has a tank that holds 50 litres of petrol. At a petrol station she adds 37.5 litres. If the tank already contains 9.02 litres, how much space is left in the tank?

```
  3 7 . 5 0
+ 0 9 . 0 2
  4 6 . 5 2
```

```
    4 9  9 1
  5 0 . 0 0
- 4 6 . 5 2
  0 3 . 4 8
```

There are 3.48 litres of space available.

DOIT!

Find an old shopping receipt and cover up the total. Then add up all the prices – do you agree with the total price?

NAILIT!

When adding and subtracting decimals, you can fill in any spaces with a '0' – this will help you to avoid making place value mistakes.

To multiply decimals, first do the calculation without any decimal point, then put a decimal point into the correct place in the answer.

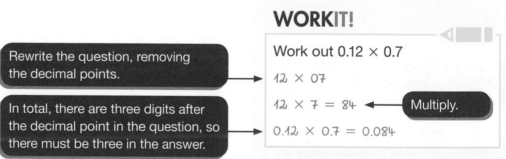

Rewrite the question, removing the decimal points.

In total, there are three digits after the decimal point in the question, so there must be three in the answer.

WORKIT!

Work out 0.12×0.7

12×07

$12 \times 7 = 84$ ← Multiply.

$0.12 \times 0.7 = 0.084$

Divide decimals in exactly the same way as you divide whole numbers – but remember to keep the decimal point in place!

STRETCHIT!

Work out
$3.2 + 7.5 \times 2$

Think about the order of operations (see page 31) – which calculation should you do first?

WORKIT!

Calculate $47.52 \div 3$

```
      1 5 . 8 4
3 ) 4 ¹7 . ²5 ¹2
```

NAILIT!

Always check your answers to see if they are sensible.

$15 \times 3 = 45$

Therefore $47.52 \div 3$ must be approximately 15.

DOIT!

Make a flashcard to remind you of the rules for working with decimals – include an example of each type.

To divide a decimal by a decimal think of the division as a fraction and change the denominator to a whole number.

WORKIT!

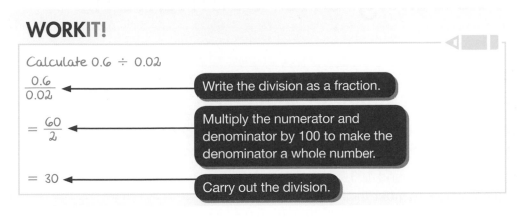

Calculate 0.6 ÷ 0.02

$\dfrac{0.6}{0.02}$ ← Write the division as a fraction.

$= \dfrac{60}{2}$ ← Multiply the numerator and denominator by 100 to make the denominator a whole number.

$= 30$ ← Carry out the division.

NAILIT!

Do not multiply your answer by 100. The fraction $\dfrac{0.6}{0.02}$ is equivalent to $\dfrac{60}{2}$ since both the numerator and denominator have been multiplied by 100.

Working with negative decimals

When calculating with negative decimals the same rules apply as when working with negative numbers.

WORKIT!

Calculate

a 3.2×-5

$3.2 \times 5 = 16$

Therefore $3.2 \times -5 = -16$

b $-1.8 + (-3.7)$

$= -1.8 - 3.7$

$= -5.5$

CHECKIT!

Do not use a calculator for these questions.

1 Work out:

  a $3.4 - 1.07$

  b $19.3 + 5.091$

  c 0.05×0.7

  d 3.21×1.9

  e $7.815 \div 5$

  f $0.18 \div 0.3$

  g $-0.3 \div 0.05$

  h 7.2×-0.7

2 Mohammad buys 24 first class stamps costing £0.64 each. How much change does he get from a £20 note?

3 In a restaurant, Erica and Freya split the bill. Erica pays twice as much as Freya. If the total bill was £82.38 how much do they each pay?

Rounding and estimation

Rounding to the nearest whole number, or to the nearest ten, hundred, and so on

We can **round** numbers to the nearest whole number, ten or hundred by thinking about which whole number, ten or hundred is nearest.

For example to round 125.7 to the nearest whole number think about which whole numbers it sits between on the number line.

125   125.7   126

It is nearer to 126, therefore 125.7 rounded to the nearest whole number is 126.

DOIT!

Round 125.7 to the:

a nearest ten

b nearest hundred.

NAILIT!

If a number sits **exactly** halfway between two numbers, round up.

172.5 sits exactly halfway between 172 and 173 so we round it up to 173.

Rounding decimals

You can round to a number of **decimal places** – this gives the number of digits after the decimal point.

You can also round to a number of **significant figures** – a significant figure is any digit after the first **non-zero** digit.

NAILIT!

Zeros **do** count as significant figures if they are **after** the first significant figure.

For example, in the number 3072 there are four significant figures: 3, 0, 7 and 2.

In the number 3879 the digit 3 is the first significant figure, the 8 is the second significant figure, the 7 is the third significant figure and the 9 is the fourth significant figure.

In the number 0.000135 the first four zeros are **not** significant figures. The digit 1 is the first significant figure, the 3 is the second significant figure and the 5 is the third significant figure.

WORKIT!

Round 3.45901 to 2 decimal places.

3.46 ◄ The last digit is rounded up when the number after it is 5 or more.

WORKIT!

1 Round 4729 to 2 significant figures.

4700 ◄ The second digit is not rounded up since the number after it is less than 5. All the other digits are replaced with zeros.

2 Round 0.00462 to 1 significant figure.

0.005 ◄ The 4 is rounded up since the number after it is 5 or more.

STRETCHIT!

Round 0.99999 to
a 1 decimal place
b 2 decimal places
c 3 decimal places.

What do you notice?

Rounding can be used to check answers in calculations, particularly when you are using a calculator as it is easy to make mistakes.

WORKIT!

Jayshuk enters 0.8×4.5 into his calculator and gets the answer 36.

Explain how he **knows** he has entered the calculation incorrectly.

0.8 is slightly less than 1, so if you multiply 0.8 by 4.5 you should get an answer smaller than 4.5.

Error intervals

An **error interval** is the range of values that a rounded number could be.

The **upper bound** is the upper limit for the largest value it could be.

The **lower bound** is the smallest value it could be.

To write an error interval, work out the upper and lower bounds.

For example, if $x = 12.3$ to 1 decimal place you can use inequality notation to write the error interval.

$12.25 \leq x < 12.35$

The smallest value that x could be is 12.25. Use the $\leq$ sign, since 12.25 would round to 12.3.

The largest value that x could be is very close to 12.35. Use the $<$ sign, since 12.35 would not round to 12.3, but anything smaller would — x could be 12.34999 for example.

WORKIT!

The width of a washing machine is given as 55 cm to the nearest cm. What is the error interval for the width of the machine?

54.5 ≤ width < 55.5

The smallest value that would round to 55 is 54.5, so this is the lower bound.

Any number that is 55.5 or larger will round up to 56 not down to 55 so this is the upper bound.

Estimation

Estimation means finding an **approximate** answer. You can find approximate answers by rounding.

WORKIT!

Petrol costs 109p per litre. Angela buys 54 litres of petrol.
Work out the approximate total cost.

100 × 50 = 5000p

5000p = £50

Round each of the values to 1 signficant figure – this makes the calculation easy!

Don't forget to give the answer in a way that makes sense – you'd never give a price as 5000p!

DOIT!

You can use a spreadsheet or an online tool to round numbers for you.

Have a go at rounding the number 3.48720912 to different numbers of significant figures or decimal places and check your answers using a spreadsheet or an online tool.

NAILIT!

Show **all** your workings – it is good practice to show all the rounded values you are using in your calculation. It also means you can check your working later.

NAILIT!

Always check your answers to see if they are sensible! Is £50 a sensible price to pay for petrol?

STRETCH IT!

Mr Appleyard measures a rectangular room to calculate the area of carpet needed. To the nearest metre, he measures the room as 6 m by 8 m.

Work out the area of carpet he should buy to ensure he has enough to cover the whole room. Is your answer an underestimate or overestimate?

CHECK IT!

1 Round each number to the level of accuracy given.

   a 0.34901   2 decimal places

   b 12.092   1 significant figure

   c 32.609   1 decimal place

   d 33 098   3 significant figures

2 The following numbers are rounded to the degree of accuracy shown. Write down the error interval for each number. Use x to represent the number in each case.

   a 200 rounded to the nearest hundred

   b 6 rounded to the nearest whole number

   c 3.2 rounded to 1 decimal place

   d 5.06 rounded to 3 significant figures

3 Round each number to 1 significant figure to find an approximate answer to:
$$\frac{31.45902}{0.491 \times 6.498}$$

4 Round 23579.82 to

   a the nearest whole number

   b the nearest ten

   c the nearest hundred

   d the nearest thousand

   e the nearest ten thousand

5 Without working out the correct answers, which of these do you **know** is false? Explain your answer.

   a $3.5 \times 7 = 24.5$

   b $18 \times 0.9 = 1.62$

   c $19 \div 0.25 = 7.6$

6 An electricity company has two different packages:

Night-time low

1.622 p per unit from 8 pm to 6 am

2.315 p per unit from 6 am to 8 pm

One tariff

1.923 p per unit all day

Tarik uses 2.32 units between 8 pm and 6 am and 7.151 units between 6 am and 8 pm.

Use this information to work out which package Tarik should choose.

Electricity charges

Delivery tariff	
Night-time low	1.622p (8 pm – 6 am)
	2.315p (6 am – 8 pm)
One tariff	1.923p
Total charges	

Converting between fractions, decimals and percentages

Percentages to fractions

'Per cent' means 'out of 100', so any percentage can be written as a fraction by writing it over 100, then cancelling the fraction down.

$$45\% = \frac{45}{100} = \frac{9}{20}$$

Percentages to decimals

Since % means 'out of 100', to change a percentage to a decimal you just need to divide by 100.

$$45\% = 45 \div 100 = 0.45$$

Decimals to percentages

To change a decimal to a percentage, multiply by 100.

$$0.63 = 63\%$$

Decimals to fractions

Remember place value?

tenths column

⇩

0.92

⇧

hundredths column

So 0.92 can be written as $\frac{92}{100} = \frac{23}{25}$

Fractions to decimals

To change a fraction to a decimal, divide the numerator by the denominator.

For example, $\frac{3}{5} = 3 \div 5$

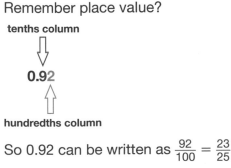

The numerator goes inside the division calculation. Add empty decimal places in case you need them.

Sometimes you will find the answer is a recurring decimal:

For example, $\frac{4}{7}$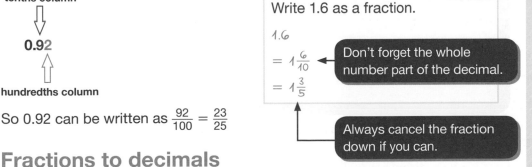

As soon as you see the number has started to repeat you know it is a recurring decimal.

WORKIT!

Write 1.6 as a fraction.

1.6

$= 1\frac{6}{10}$ — Don't forget the whole number part of the decimal.

$= 1\frac{3}{5}$

Always cancel the fraction down if you can.

NAILIT!

Not sure what to use as the denominator of the fraction?

Look at where the last digit is within the decimal – this will tell you whether the denominator is 10, 100 or 1000 and so on.

For example, in 0.456 the last digit (6) is in the thousandths column, so $0.456 = \frac{456}{1000}$

Don't forget to cancel down if you can!

STRETCHIT!

Work out what $\frac{1}{9}$, $\frac{2}{9}$ and $\frac{3}{9}$ are as decimals. Predict the decimal values of $\frac{4}{9}$, $\frac{5}{9}$ and so on. Can you explain what is happening?

To indicate that a series of digits are recurring, place a dot over the first and last recurring digits:

$$\frac{4}{7} = 0.\dot{5}7142\dot{8} = 0.571428571428571428\ldots$$

Fractions to percentages

To change a fraction to a percentage, first change it to a decimal. Once you've written the fraction as a decimal, multiply by 100.

For example, $\frac{3}{5} = 0.6 = 60\%$

DOIT!

Cut out six pieces of card the size of playing cards. On one side write the six titles:

Percentages to fractions   *Decimals to fractions*

Fractions to decimals   *Percentages to decimals*

Decimals to percentages   *Fractions to percentages*

On the back of each card, write in your own words how you would convert them.

Spread the cards out in front of you with the titles showing. Test yourself to check you know how to carry out all the conversions – you could even play with a friend and make it a competition!

SNAPIT!

Fractions to percentages

$\frac{1}{2} = 0.5$	$= 50\%$
$\frac{1}{4} = 0.25$	$= 25\%$
$\frac{3}{4} = 0.75$	$= 75\%$
$\frac{1}{3} = 0.\dot{3}$	$= 33.\dot{3}\%$
$\frac{2}{3} = 0.\dot{6}$	$= 66.\dot{6}\%$
$\frac{1}{10} = 0.1$	$= 10\%$
$\frac{1}{5} = 0.2$	$= 20\%$
$\frac{1}{8} = 0.125$	$= 12.5\%$

NAILIT!

To compare fractions, decimals and percentages, write them all in the same form first.

CHECKIT!

Complete these questions without using a calculator.

1 Convert to fractions

    a 0.32   **c** 33%

    b 1.24   **d** 95%

2 Convert to decimals

    a $\frac{5}{12}$ **c** 49%   **e** $\frac{3}{7}$

    b $\frac{3}{8}$ **d** 18.5%

3 Convert to percentages

    a 0.91   **c** $\frac{4}{5}$

    b 0.3   **d** $\frac{9}{15}$

4 What percentage of the shape is shaded?

5 Place these quantities in order of size, from smallest to largest.

    0.35 $\frac{2}{5}$ 30%

6 Amy correctly answered 15 out of 20 questions in a test. Rudi achieved 78%. Whose mark was higher? Show your workings.

Ordering fractions, decimals and percentages

Ordering fractions

To compare the size of fractions give them a **common denominator**.

To find the common denominator find the **lowest common multiple** of the denominators.

> See page 10 for more on lowest common multiples.

WORKIT!

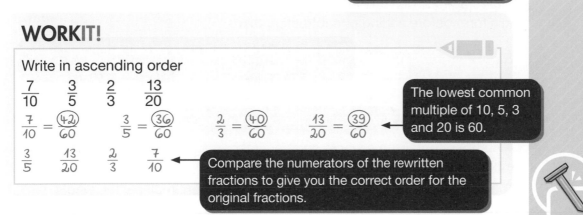

Write in ascending order

$$\frac{7}{10} \quad \frac{3}{5} \quad \frac{2}{3} \quad \frac{13}{20}$$

$$\frac{7}{10} = \frac{42}{60} \qquad \frac{3}{5} = \frac{36}{60} \qquad \frac{2}{3} = \frac{40}{60} \qquad \frac{13}{20} = \frac{39}{60}$$

$$\frac{3}{5} \quad \frac{13}{20} \quad \frac{2}{3} \quad \frac{7}{10}$$

> The lowest common multiple of 10, 5, 3 and 20 is 60.

> Compare the numerators of the rewritten fractions to give you the correct order for the original fractions.

Ordering fractions, decimals and percentages

Follow these steps when ordering fractions, decimals and percentages.

1. Split into < 0 and > 0.

2. Convert all the numbers into the same type of number, either fractions, decimals or percentages, to make them easy to compare.

> < 0: -5.4, $-\frac{1}{2}$
> > 0: $\frac{3}{4}$, 17%, 0.4
> Writing all these numbers as decimals gives: -5.4, -0.5, 0.75, 0.17, 0.4

WORKIT!

Write in **ascending** order -5.4, $\frac{3}{4}$, 17%, $-\frac{1}{2}$, 0.4.

In ascending order:

$$-5.4, \; -\frac{1}{2}, \; 17\%, \; 0.4, \; \frac{3}{4}$$

> Don't forget to write the original numbers in order.

NAILIT!

Use your common sense: if you know that -5.4 is smaller than $-\frac{1}{2}$ you don't need to convert $-\frac{1}{2}$ to a decimal.

CHECKIT!

1. Write in **descending** order

   $$\frac{1}{3} \qquad \frac{3}{8} \qquad \frac{7}{12}$$

2. If these numbers were written in order, which would be in the middle?

   $$-2.2 \quad 7 \quad \frac{1}{5} \quad -\frac{1}{10} \quad 15\% \quad 1\% \quad 0.1$$

3. Jenny says, 'If the numerator of a fraction is less than half of the denominator I know the number is smaller than 0.5'.

   Is she correct? Give a reason for your answer.

Calculating with fractions

Adding and subtracting fractions

In order to be able to add or subtract fractions you need them to have a common denominator.

If fractions already have a common denominator you can add or subtract them by adding or subtracting the numerators.

$$\frac{3}{5} + \frac{1}{5} = \frac{3 + 1}{5} = \frac{4}{5} \qquad\qquad \frac{9}{11} - \frac{7}{11} = \frac{9 - 7}{11} = \frac{2}{11}$$

If the fractions don't have a common denominator you need to get them in this form before adding or subtracting them.

$$\frac{2}{3} + \frac{1}{12} = \frac{8}{12} + \frac{1}{12}$$

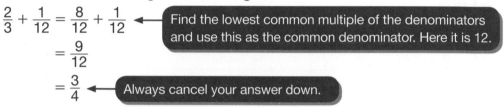 Find the lowest common multiple of the denominators and use this as the common denominator. Here it is 12.

$$= \frac{9}{12}$$

$$= \frac{3}{4}$$

Always cancel your answer down.

A number containing a whole number (integer) and a fraction is called a **mixed number**.

To convert a mixed number to an improper fraction, multiply the whole number part by the denominator and add this to the existing numerator.

For example in $2\frac{3}{5}$ the whole number part is worth $\frac{(2 \times 5)}{5}$ so:

$$2\frac{3}{5} = \frac{10 + 3}{5} = \frac{13}{5}$$

A fraction whose numerator is larger than the denominator is called an **improper fraction**.

To convert an improper fraction to a mixed number, divide the numerator by the denominator to give a whole number and a remainder. Write down the integer first, then write the remainder over the original denominator.

For example, to write $\frac{23}{20}$ as a mixed number, divide 23 by 20 to give 1 remainder 3: write 1 and put the 3 over 20.

$$\frac{23}{20} = 1\frac{3}{20}$$

DO IT!

Draw a flow diagram to show how to add and subtract fractions.

WORKIT!

To add or subtract mixed numbers, first write them as improper fractions.

Work out $2\frac{2}{5} - 1\frac{1}{4}$

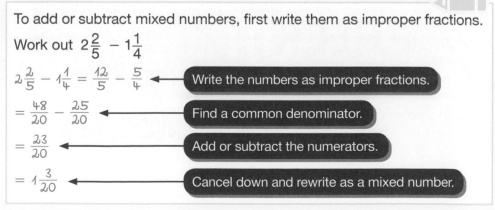

$$2\frac{2}{5} - 1\frac{1}{4} = \frac{12}{5} - \frac{5}{4}$$

Write the numbers as improper fractions.

$$= \frac{48}{20} - \frac{25}{20}$$

Find a common denominator.

$$= \frac{23}{20}$$

Add or subtract the numerators.

$$= 1\frac{3}{20}$$

Cancel down and rewrite as a mixed number.

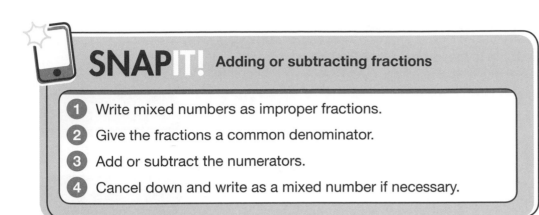

SNAP IT!   Adding or subtracting fractions

1. Write mixed numbers as improper fractions.
2. Give the fractions a common denominator.
3. Add or subtract the numerators.
4. Cancel down and write as a mixed number if necessary.

STRETCH IT!

Do you always need to write the fractions as improper fractions before adding?

E.g. what is the best way to work out $1\frac{3}{5} + 2\frac{1}{4}$?

Multiplying fractions

To multiply fractions, work out: $\dfrac{\text{numerator} \times \text{numerator}}{\text{denominator} \times \text{denominator}}$

To multiply mixed numbers:

1. First write them as improper fractions.

$$1\frac{2}{3} \times \frac{4}{5} = \frac{5}{3} \times \frac{4}{5}$$

2. Multiply the numerators and the denominators.

$$= \frac{5 \times 4}{3 \times 5} = \frac{20}{15}$$

3. Cancel down and write as a mixed number.

$$= \frac{4}{3} = 1\frac{1}{3}$$

WORKIT!

Calculate $\dfrac{3}{4} \times \dfrac{6}{7}$

$$\frac{3 \times 6}{4 \times 7} = \frac{18}{28}$$

$$= \frac{9}{14} \longleftarrow \boxed{\text{Make sure you cancel down.}}$$

Cross cancelling

When multiplying fractions you can simplify the calculation by cross cancelling first. This means dividing the numerator of one fraction and the denominator of the other by the same amount.

WORKIT!

Work out $\dfrac{3}{20} \times \dfrac{10}{21}$

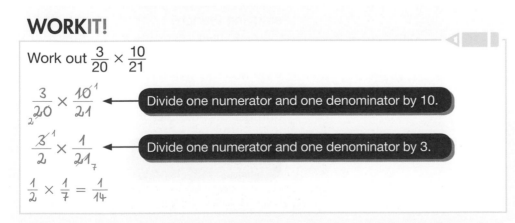

$$\frac{3}{20} \times \frac{10^{1}}{21} \longleftarrow \boxed{\text{Divide one numerator and one denominator by 10.}}$$

$$\frac{3^{1}}{2} \times \frac{1}{21_{7}} \longleftarrow \boxed{\text{Divide one numerator and one denominator by 3.}}$$

$$\frac{1}{2} \times \frac{1}{7} = \frac{1}{14}$$

DO IT!

Multiply $\dfrac{3}{20} \times \dfrac{10}{21}$ without cross cancelling, then cancel your answer down. Now try again, but cross cancel first – you should get the same answer. Which method was easier?

DOIT!

Write down two different mixed numbers.

For example, $2\frac{3}{5}$ and $1\frac{2}{3}$

Practise adding, subtracting, multiplying and dividing them. You can check your answers using the fraction key on your calculator.

NAILIT!

Multiplying by $\frac{1}{a}$ is the same as dividing by a.

For example, multiplying by $\frac{1}{5}$ is the same as dividing by 5.

Dividing fractions

Dividing is the inverse of multiplying, so to divide fractions flip the second fraction and multiply.

To divide mixed numbers:

1 First write them as improper fractions.

$$2\frac{2}{3} \div \frac{3}{4} = \frac{8}{3} \div \frac{3}{4}$$

2 Flip the second fraction upside down and multiply.

$$= \frac{8}{3} \times \frac{4}{3} = \frac{32}{9}$$

3 Cancel down if necessary and write as a mixed number.

$$= 3\frac{5}{9}$$

WORKIT!

Calculate $\frac{3}{4} \div \frac{1}{2}$

$$= \frac{3}{4} \times \frac{2}{1}$$ ◄ Flip the 2nd fraction and multiply.

$$= \frac{6}{4}$$

$$= 1\frac{2}{4}$$

$$= 1\frac{1}{2}$$ ◄ Write your answer as a mixed number.

Finding fractions of quantities

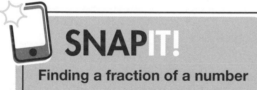

SNAPIT!

Finding a fraction of a number

To find a fraction of a number, divide by the denominator, then multiply by the numerator.

WORKIT!

Find $\frac{2}{3}$ of £12.45.

Divide by the denominator, 3.

£12.45 ÷ 3 = £4.15

£4.15 × 2 = £8.30 ◄

Then multiply by the numerator, 2.

CHECKIT!

1 Work out

a $2\frac{3}{8} - \frac{3}{4}$

b $\frac{15}{17} \times \frac{2}{5}$

c $\frac{1}{7} \times 3\frac{1}{3}$

d $2\frac{2}{5} + 5\frac{3}{4}$

e $\frac{1}{5} \div 2\frac{1}{2}$

2 Find

a $\frac{2}{5}$ of 30

b $\frac{7}{8}$ of £40

c $\frac{4}{9}$ of 1818 mm ◄ Here, 'of' means the same as ×.

3 In a class of 35 students, $\frac{3}{7}$ are studying French. How many students are not studying French?

4 Adrian thinks of a number. It is smaller than 50, and $\frac{2}{5}$ of it is a whole number greater than 12. What is the smallest number it could be?

Percentages

Calculating a percentage of a value

You need to be able to find a percentage with and without a calculator. Remember that per cent (%) means 'out of 100'.

To find a percentage without a calculator, work out simpler percentages and add them together.

For example, to find 75% you could work out 50% and 25% and add them together.

SNAP IT!
Percentages without a calculator

You can work out some percentages without a calculator:

50% – halve it

25% – halve it then halve it again

10% – divide by 10

5% – find half of 10%

1% – divide by 100

DO IT!

Which simpler percentages would you work out to find each of these?

30%   45%   80%

You may find there is more than one method!

WORKIT!

Find 37% of £16.00.

£16.00 ÷ 10 = £1.60   ← First find 10%.

£1.60 ÷ 2 = £0.80   ← Then find 5%.

£16.00 ÷ 100 = £0.16   ← Next find 1%.

37% = (3 × 10%) + 5% + (2 × 1%)

= (3 × £1.60) + £0.80 + (2 × £0.16)

= £4.80 + £0.80 + £0.32

= £5.92

SNAP IT!
Percentages with a calculator

To find a percentage with a calculator, write it as a decimal, then multiply.

For example, to find 37%, multiply by 0.37.

Increasing or decreasing by a percentage

There are two ways to increase or decrease by a given percentage.

Method 1

Work out the increase and add it to the original value, or work out the decrease and subtract it from the original value.

WORKIT!

In a sale, a tablet is reduced by 20%. The original price of the tablet was £399. Work out the sale price.

£399 × 0.20 = £79.80   ← First calculate 20%.

£399 − £79.80 = £319.20

Since the price is being reduced, subtract from the original value.

NAIL IT!

Take care! One of the most common mistakes with percentage increase or decrease is forgetting to add or subtract from the original price.

Method 2

Work out the multiplier. For example, if a value is increased by 20% then you need to find 120% of the original value.

120% = 1.2 so multiply the value by 1.2

If the value is decreased by 20%, then you need to find 80% of the original value.

80% = 0.8 so multiply the value by 0.8

WORKIT!

A shop reduces its prices by 10%. If the original price of a laptop computer is £380, work out the sale price.

0.9 × 380 = £342 ◄─── If the price is reduced by 10% you need to find 90% of the original price, so multiply by 0.9.

DOIT!

Draw a flow diagram to show how to increase or decrease by a given percentage.

CHECKIT!

1 Without a calculator, work out

   a 10% of 18 cm

   b 25% of £1.20

   c 2% of 200 ml

2 Increase

   a 30 by 10%

   b 500 by 8%

   c £91 by 12%

3 Decrease

   a 600 by 20%   **c** £18 by 19%

   b 140 by 5%

4 In 2016, a college had 2800 students. By 2017, the number of students had increased by 9%. How many students attended the college in 2017?

5 A car loses 35% of its value in the first year. What is the value of a 1-year-old car that originally cost £22 000?

Order of operations

When performing calculations, you must follow the **order of operations** — the priority in which you work things out.

> When the numerator and the denominator of a fraction are separate calculations, work them out separately before dividing.

WORKIT!

Calculate $\dfrac{3 \times 4^2 + 12}{10 - (1 + 1)}$

$3 \times 4^2 + 12$ ← Indices first.

$= 3 \times 16 + 12$ ← Multiplication.

$= 48 + 12$ ← Addition.

$= 60$

$10 - (1 + 1)$ ← Brackets.

$= 10 - 2$ ← Subtraction.

$= 8$

$\dfrac{60}{8} = 7\dfrac{1}{2}$ ← Division.

SNAPIT!
BIDMAS

Use **BIDMAS** to remember the correct order of operations.

Brackets
Indices
Division
Multiplication
Addition
Subtraction

NAILIT!

Although we call it BIDMAS, multiplication and division can happen in any order. So can addition and subtraction.

NAILIT!

When interpreting your calculator's answer, think about what it means and the units you are using. For example, 0.6 pounds (£) should be written down as 60p or £0.60, depending what the question asks for.

Using a calculator

Your calculator will follow the order of operations as long as you type in the correct calculation.

SNAPIT!   Calculator keys

( ) brackets	√ find the square root of a number	ANS use the previous answer in a calculation
x^2 square a number	$\sqrt[\square]{x}$ find a root of a number	S-D change a number from an exact value to a decimal
$x^\square$ raise a number to a power	(−) enter a negative number	$\dfrac{a}{b}$ or $\dfrac{\square}{\square}$ enter a fraction

> When you use a calculator to work out a problem with more than one step, make sure you use all the digits in your answer in the next part of the calculation.

CHECKIT!

1 Without a calculator, work out:

   a $3 - 4 + 8$ **c** $(5 - 6)^2 - 7 \times 2$

   b $0.9 + 3.2 - \sqrt{35 + 1}$

2 Use a calculator to work out: $\dfrac{3.2^3 - \sqrt{1.4 - 0.99}}{2 - 8.6 \div 9.2}$

   Give your answer to 2 significant figures.

3 Put a pair of brackets into this calculation to make it as large as possible.

   $8 - 3 + 5 \times 4$

DOIT!

Without a calculator, work out

$2 + 5.5 \times 3 - 16 \div 2^2 - (3.1 + 0.9)$

Now check it on your calculator — make sure you type it in exactly as it is written!

Exact solutions

NAILIT!

Most scientific calculators have a button that converts between an exact solution and a decimal solution. On some calculators it is the S-D key. Work out which it is on your calculator.

You must be able to work out exact solutions without using a calculator. To give an exact answer you may need to include π or fractions or square roots as part of the answer.

When adding multiples of π, treat it in the same way as simplifying an algebraic expression.

For example: $3\pi + 7\pi = 10\pi$

WORKIT!

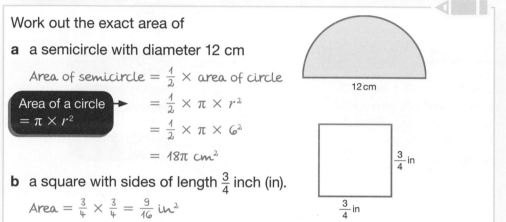

Work out the exact area of

a   a semicircle with diameter 12 cm

Area of semicircle $= \frac{1}{2} \times$ area of circle

Area of a circle $= \pi \times r^2$

$= \frac{1}{2} \times \pi \times r^2$

$= \frac{1}{2} \times \pi \times 6^2$

$= 18\pi \ cm^2$

b   a square with sides of length $\frac{3}{4}$ inch (in).

Area $= \frac{3}{4} \times \frac{3}{4} = \frac{9}{16} \ in^2$

DOIT!

Type into your calculator the calculation $3 \times 4\pi$.

What answer does it give you? Convert this answer to a decimal.

NAILIT!

Always sketch a shape if you are not given a picture, labelling any lengths you know. This will help you visualise the problem.

✓ CHECKIT!

1 Write as a single multiple of π

   a $2\pi - \pi$ **c** $\frac{1}{2}\pi + 2\pi$

   b $12\pi + 24\pi$

2 Find the exact solution to:

   a $(3 + 4) \times \pi$ **b** $\frac{1}{8}\pi + \frac{1}{2}\pi$

3 Work out the exact area and perimeter of this shape.

   $\frac{3}{4}$ cm

   $\frac{2}{7}$ cm

4 Work out the exact value of each of these.

   a The circumference of a circle with radius 9 cm.

   b The area of a circle with radius 12 cm.

5 A square has perimeter equal to the circumference of a circle with radius 1 cm. Work out the exact length of one of the sides of the square.

Indices and roots

Indices

Indices are sometimes called 'powers'. An index tells you how many times to multiply a number by itself.

For example, 5^4 means $5 \times 5 \times 5 \times 5$.

Reciprocals

The reciprocal of a number is 1 divided by that number.
The reciprocal of 4 is $\frac{1}{4}$. It can be written as 4^{-1}.

A number multiplied by its reciprocal is 1. For example, $3^{-1} \times 3 = 1$.

The inverse of multiplying by a number is to multiply by the reciprocal.
For example, the inverse of multiplying by 4 is multiplying by $\frac{1}{4}$.

A negative index means 1 divided by that number (the reciprocal of the number). For example, 3^{-2} means the reciprocal of $3^2 = \frac{1}{3^2} = \frac{1}{9}$.

WORKIT!

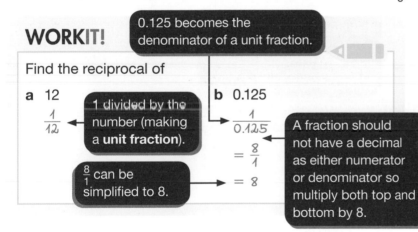

Find the reciprocal of

a 12

$\frac{1}{12}$ ← 1 divided by the number (making a **unit fraction**).

$\frac{8}{1}$ can be simplified to 8.

b 0.125

0.125 becomes the denominator of a unit fraction.

$\frac{1}{0.125}$

$= \frac{8}{1}$

$= 8$

A fraction should not have a decimal as either numerator or denominator so multiply both top and bottom by 8.

Special indices and laws of indices

Any number to the power of 1 is itself. $7^1 = 7$

1 to any power is 1. $1^{12} = 1$

Any number to the power of 0 is 1. $3^0 = 1$

The **laws of indices** (see Snap it!) can be used to simplify expressions.

WORKIT!

a Simplify the expression $4^2 \times 4^{-5}$, leaving your answer in index form.

$4^2 \times 4^{-5} = 4^{(2 + -5)} = 4^{-3}$

b Work out the value of $\frac{(3^7)^2}{3^{11}}$

$\frac{(3^7)^2}{3^{11}} = \frac{3^{7 \times 2}}{3^{11}} = \frac{3^{14}}{3^{11}} = 3^{14-11} = 3^3 = 3 \times 3 \times 3 = 27$

WORKIT!

1 Write the following in index form

a $5 \times 5 \times 5 \times 5 \times 5 \times 5$

5^6

b 9×9

9^2

2 Work out the value of

a 3^2

$3 \times 3 = 9$

b 3^4

$3 \times 3 \times 3 \times 3 = 81$

c $(-1)^3$

$-1 \times -1 \times -1 = -1$

SNAP IT!   Special indices

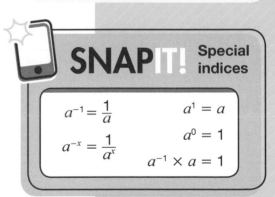

$a^{-1} = \frac{1}{a}$ $a^1 = a$

$a^{-x} = \frac{1}{a^x}$ $a^0 = 1$

$a^{-1} \times a = 1$

SNAP IT!

Laws of indices

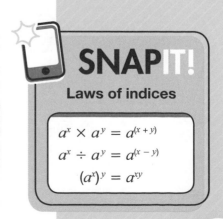

$a^x \times a^y = a^{(x + y)}$

$a^x \div a^y = a^{(x - y)}$

$(a^x)^y = a^{xy}$

Square roots

The square root is the inverse of squaring a number. The square root sign is $\sqrt{}$

$\sqrt{49}$ means 'the number that multiplies by itself to give 49'.

$\sqrt{49} = 7$ or -7 ◄── Two negatives make a positive when multiplied.

If the exact root of a number cannot be found then it can be left written as a root. This is called a **surd** – for example, $\sqrt{7}$.

You can **estimate the square root** of a more complex calculation by rounding the numbers to a sensible degree of accuracy.

NAIL IT!

When you find the square root of a number there are two solutions, negative and positive.

For example,
$-7 \times -7 = 49$ **and**
$7 \times 7 = 49$

STRETCH IT!

Other roots

The sign for other roots is $\sqrt[a]{}$

$\sqrt[3]{8}$ means 'the number that multiplies by itself 3 times to make 8'.

$\sqrt[4]{10\,000}$ means 'the number that multiplies by itself 4 times to make 10 000'.

WORKIT!

Estimate the value of
$$\sqrt{\frac{0.49 \times 823}{98.8}}$$
Rounding each number to 1 significant figure gives:

$$\sqrt{\frac{0.49 \times 823}{98.8}} \approx \sqrt{\frac{0.5 \times 800}{100}}$$

$$= \sqrt{\frac{400}{100}}$$

$$= \sqrt{4}$$

$$= 2 \text{ or } -2$$

The 'wavy' equals sign ($\approx$) means 'approximately equal to'. Use it when you are approximating a value.

WORKIT!

1 Work out $\sqrt{64}$

± 8

2 Calculate

a $\sqrt[3]{8}$

$2 \times 2 \times 2 = 8$
So $\sqrt[3]{8} = 2$

b $\sqrt[10]{1}$

$1 \times 1 \times 1 \times 1 \times 1 \times 1$
$\times 1 \times 1 \times 1 \times 1 = 1$
So $\sqrt[10]{1} = 1$ or -1

The square root of a larger number can be found by writing it as a product of prime factors. For example, to find the square root of 225:

Write as a product of prime factors:

$225 = 3^2 \times 5^2$.

Remember that the square root is the inverse of squaring, so: $\sqrt{225} = \sqrt{3^2 \times 5^2} = 3 \times 5 = 15$.

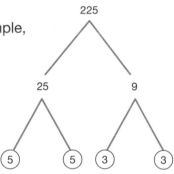

NAIL IT!

Any root of 1 is 1.

For example, $\sqrt[15]{1} = 1$.

CHECKIT!

1 Find the reciprocal of

   a 3   **b** 0.4   **c** 0.9

2 Write these six numbers in order of size, from smallest to largest.

   3^2 1^3 $\sqrt[3]{27}$ $\sqrt[3]{8}$ 3.7 $\dfrac{1}{12}$

3 Work out

   a $(-2)^3$ **b** 1^{12} **c** $(-9)^2$ **d** 17^0

4 What is the value of

   a 4^{-1} **c** 1^{-4}

   b 7^{-2} **d** the reciprocal of 3?

5 Simplify the expression $\dfrac{5^9}{5^2 \times 5^3}$ leaving your answer in index form.

6 Work out an approximate value of:
$(0.01 \times 798)^2$

Standard form

SNAPIT! **Standard form**

To write 0.00000567 in standard form:

1 Put the decimal point after the first **non-zero** digit.

5.67

2 Work out how many times you have multiplied or divided by 10 and use this as the power of 10.

6 times, so use 10^6

3 If the original number is very large then the index is positive. If it is very small then the index is negative.

5.67×10^{-6}

Standard form is used to write very large and very small numbers. Numbers in standard form are written in two parts, a numerical value (1 or more, but less than 10) and a power of 10. For example:

5.4×10^3.

The number can be written in 'ordinary form' by multiplying out.

$5.4 \times 10^3 = 5.4 \times 10 \times 10 \times 10$
$= 5400$

If the power of 10 is negative you can read it as dividing by the power of 10.

$3.2 \times 10^{-3} = 3.2 \div (10 \times 10 \times 10)$
$= 0.0032$

Calculations involving standard form

To multiply or divide numbers written in standard form use the laws of indices.

For example: $(3.2 \times 10^5) \times (5 \times 10^9)$ can be rewritten as 1.6×10^{15}.

$3.2 \times 5 \times 10^5 \times 10^9 = 16 \times 10^{14}$

$16 \times 10^{14} = 1.6 \times 10^{15}$

Combine the powers of 10.

Divide 16 by 10 (to give 1.6) and multiply 10^{14} by 10 (to give 10^{15}), so that your answer is written correctly in standard form.

WORKIT!

a Work out $(2.5 \times 10^3) + (4.2 \times 10^5)$.

$(2.5 \times 10 \times 10 \times 10) + (4.2 \times 10 \times 10 \times 10 \times 10 \times 10)$
$= 2500 + 420\,000 = 422\,500$

b Write 9.5×10^{-5} as an ordinary number.

$9.5 \div (10 \times 10 \times 10 \times 10 \times 10) = 0.000095$

CHECKIT!

When you've finished these questions, check all your answers on your calculator.

1 Write these numbers as ordinary numbers.
   a 4.5×10^7 b 9.1×10^{-2}

2 Write these numbers in standard form.
   a 645 000 000   b 0.000000079

3 Work out the value of $(3.5 \times 10^5) - (4.2 \times 10^3)$

4 Write these four numbers in order of size from smallest to largest.

3.2×10^2 3.1×10^{-2}
3.09×10 $3 + (2.1 \times 10^2)$

5 The speed of light is approximately 300 000 000 m/s. Write this in standard form.

6 A piece of A4 paper is 1.1×10^{-4} metres thick. Work out the height of a stack of 200 sheets.

7 Check all your answers to questions 1 – 6 on your calculator.

DOIT!

You can use your calculator to carry out calculations involving standard form.

To work out:
$3.25 \times 10^5 + 5.1 \times 10^{-3}$

use the button on your calculator that looks like: 10^n or 10

Work out the solution to this calculation on your calculator.

Listing strategies

STRETCHIT!

You could alternatively start by considering each colour at a time. How would the list look then?

To work out the total number of combinations of different events you must use strategies to avoid missing out or repeating outcomes.

For example, if a shop sells T-shirts in three different sizes (small, medium and large) and three different colours (red, green and blue), you could consider the different options for one size at a time, as shown below.

small + red	small + green	small + blue
medium + red	medium + green	medium + blue
large + red	large + green	large + blue

Sample space diagrams

You can draw up a sample space diagram to help you list the possible outcomes of an event.

For the shop selling T-shirts, draw up a table with the size options horizontally and the colour options vertically.

	small	medium	large
red	red + small	red + medium	red + large
green	green + small	green + medium	green + large
blue	blue + small	blue + medium	blue + large

Drawing up a sample space can help prevent you missing out any combinations.

See page 185 for more on sample space diagrams.

DOIT!

A dice with numbers 1 to 6 and a coin with sides showing heads and tails are thrown.

List all the different possible outcomes.

Then draw up a sample space to show all the possible outcomes. Which method works better?

CHECKIT!

1 List all the different three-digit numbers that can be made using the digits 1, 2 and 3.

2 List all the different numbers smaller than 500 that can be made using the digits 4, 6 and 9.

3 A factory manufactures shampoo in three different sizes: small, medium and large. They make four different types of shampoo: almond, briar rose, chamomile and detox. List all the different products the factory produces.

You may **not** use a calculator for these questions.

1 34 592 tickets are sold by a train company one morning. Of those, 21 298 are bought online and the rest are purchased at the station. How many are purchased at the station?

2 Write down a prime number and a factor of 12 that add up to 13.

3 Write 620 as a product of prime factors using indices.

4 Find the highest common factor of 18, 36 and 40.

5 Write these numbers in ascending order:

-3.2 -11.5 1.4 -8.3 -3.5

6 Calculate
  a £32.99 + £18.74 **b** £54.99 ÷ 3

7 Work out
  a 23×0.14 **d** $-13.5 + 8.7$
  b 27×149 **e** $-1.2 \div 0.3$
  c -0.3×0.4

8 Round each of the numbers in the calculation to 2 significant figures. Hence estimate the solution.

$30.45 \times \sqrt{15.92} + 17.395$

9 Work out $\frac{1}{3}$ of 81.

10 Work out 345 ÷ 11, giving your answer as a mixed number.

11 **a** Write $\frac{3}{8}$ as a decimal.
  b Write 0.7 as a percentage.

12 Give your answer to each of these calculations in its simplest form.
  a Write 70% as a fraction.
  b Write 0.8 as a fraction.

13 Which of these fractions is larger than $\frac{1}{4}$?
  $\frac{1}{2}$ $\frac{2}{7}$ $\frac{3}{11}$ $\frac{2}{5}$

14 Give your answer to each of these calculations in its simplest form.
  a $\frac{3}{5} + \frac{1}{7}$ **b** $2\frac{1}{5} - \frac{7}{10}$ **c** $\frac{2}{3} \div \frac{4}{9}$

15 Which of these is larger?
  $0.25 - 0.07$ or $\frac{2}{3} - \frac{1}{2}$

16 Work out $\frac{3}{5} \times 1\frac{1}{4}$.

17 Write 0.045 as a fraction. Give your answer in its simplest form.

18 Find 25% of £8.60.

19 Work out
  a $(-3)^2$ **b** $\sqrt[3]{125}$

20 Write each of the following in standard form
  a 3 400 000 000 **b** 0.000000304

21 The length of a journey is measured as 37.6 miles to the nearest tenth of a mile. Write down the error interval for the journey.

22 Here are three cards.
  a Write down the largest two-digit multiple of 3 that can be made using the cards. Each card may be used only once. [2] [5] [1]
  b Write down all the two-digit numbers that can be made using these cards.

23 A unit of electricity costs 9.086p. On average a family uses 242.6 units a month.
  a Work out an estimate for the cost of electricity for a year.
  b Is your answer an overestimate or an underestimate? Give a reason for your answer.

24 At a school of 600 students, 40% of the students cycle to school and $\frac{1}{5}$ walk. The rest travel by bus. How many travel by bus?

25 Rhiannon's result in a test is less than 50%. Her mark is a multiple of 5. She answered more than $\frac{1}{3}$ of the questions correctly. What is the smallest value her mark could be?

26 Barbara says, 'The sum of three different prime numbers will always be odd.'
  Is she correct? Explain your answer.

You **may** use a calculator for these questions.

27 In a sale, a musical instrument that normally costs £349 is reduced by 20%. What is the new price of the instrument?

28 Round 3.05092 to
  a 1 decimal place **b** 3 significant figures.

29 Round 324 567 to
  a the nearest thousand **b** 2 significant figures.

30 Find the value of 3^6.

31 Four people share the cost of an outing equally between them. The costs are:

Group travel card	£26.25
Picnic	£18.23
4 tickets at	£5.50 each

  How much does each person pay?

32 19% of people in a town own a cat. The population is 18000. How many people own a cat?

33 A tablet cost £102.30 in 2016. The price went up by 10% in 2017. What was the price in 2017?

Algebra

Understanding expressions, equations, formulae and identities

DO IT!

Draw a poster to highlight the differences between expressions, equations, formulae and identities. Use your own words and include an example of each.

In algebra you use letters to represent unknown numbers.

$4a + 2b$ is an **expression**.

It does not have an equals (=) sign. The parts that are separated by + or − are called **terms**. In $4a + 2b$ the terms are $4a$ and $2b$. An expression can contain letter terms and/or number terms.

$2x + 2 = 8$ is an **equation**.

It has an equals sign. It contains letter terms and numbers. You can solve an equation to find the value of the letter. An equation is only true for certain values of the letter. Here, the value of x is 3.

$A = lw$ is a **formula** (plural: **formulae**).

It has an equals sign. The letters represent different quantities. The letters are **variables**, as their values can vary. You can use a formula to calculate one variable if you know the other variables. For example, you can use $A = lw$ to find A if you know l and w.

$\frac{10x}{2} \equiv 5x$ is an **identity** (plural: **identities**).

An identity is true for all values of the letters. Here, the two sides of $\frac{10x}{2} \equiv 5x$ are equal for all values of x. An identity can be written with the identity symbol '≡' or with an equals sign. (See page 69 for more on identities.)

WORKIT!

Write down whether each of these is an expression, an equation, a formula or an identity.

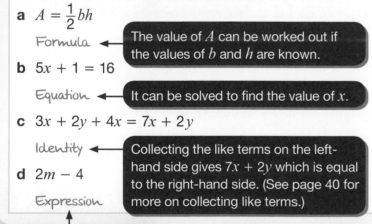

a $A = \frac{1}{2}bh$

Formula — The value of A can be worked out if the values of b and h are known.

b $5x + 1 = 16$

Equation — It can be solved to find the value of x.

c $3x + 2y + 4x = 7x + 2y$

Identity — Collecting the like terms on the left-hand side gives $7x + 2y$ which is equal to the right-hand side. (See page 40 for more on collecting like terms.)

d $2m - 4$

Expression

It does not have an equals sign.

CHECKIT!

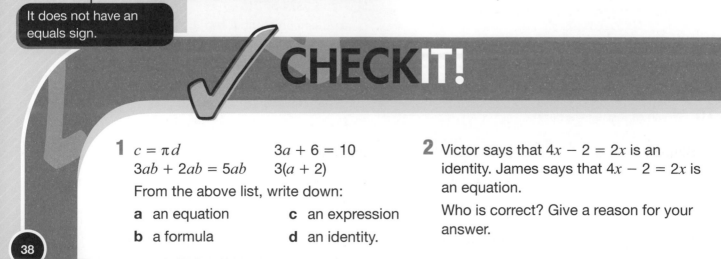

1

$c = \pi d$ $3a + 6 = 10$

$3ab + 2ab = 5ab$ $3(a + 2)$

From the above list, write down:

a an equation   **c** an expression

b a formula   **d** an identity.

2 Victor says that $4x - 2 = 2x$ is an identity. James says that $4x - 2 = 2x$ is an equation.

Who is correct? Give a reason for your answer.

Simplifying expressions

STRETCHIT!

When three algebraic expressions are multiplied together, the answer is $12t^3$. Write down the possible multiplied expressions.

For an unknown number x, the **expression** $2x + 3$ means '2 lots of x with 3 added' or 'add 3 to 2 lots of x'. '2 lots of x' is the same as '$2 \times x$' or '$x + x$' and is written as $2x$.

Expressions that involve $\times$ and $\div$ should be simplified where possible.

Multiplying expressions

To multiply together two or more algebraic expressions, first multiply any number parts and then multiply the letters.

SNAPIT! Algebraic notation

$3a$ is used for $3 \times a$
ab is used for $a \times b$
a^2 is used for $a \times a$
a^3 is used for $a \times a \times a$
a^2b is used for $a \times a \times b$
$\dfrac{a}{b}$ is used for $a \div b$

WORKIT!

Simplify:

a $a \times a$

a^2

Use indices to show a letter multiplied by itself. a^2 is read as 'a squared'.

b $b \times b \times b$

b^3

Use indices to show the same letter multiplied together three times. b^3 is read as 'b cubed'.

c $2 \times j \times k \times 6$

$12jk$

Multiply the numbers together:
$2 \times 6 = 12$
Multiply the letters together by writing them next to each other:
$j \times k = jk$
Write the number in front of the letters.

d $2d \times 4d$

$8d^2$

Dividing expressions

Dividing algebraic expressions is similar to cancelling fractions. If necessary, write the division as a fraction first. Cancel any number parts where possible and then cancel any letters where possible.

Write the division as a fraction.
Cancel the number parts by dividing the top and bottom by 2.

WORKIT!

Simplify:

a $6a \div 2$

$\dfrac{\overset{3}{\cancel{6}}a}{\cancel{2}} = 3a$

b $12xy \div 4y$

$\dfrac{\overset{3}{\cancel{12}}x\cancel{y}}{\cancel{4}\cancel{y}} = 3x$

This is the same as $\dfrac{12 \times x \times y}{4 \times y}$.
Cancel the number parts by dividing the top and bottom by 4.
y appears on the top and bottom, so cancel.

CHECKIT!

1 Simplify:

a $p \times p \times p$

b $4 \times b \times c \times 7$

c $4a \times 3b$

d $5x \times 4x$

e $2g \times (-4g)$

f $2p \times 3q \times r$

2 Simplify:

a $10x \div 2$

b $\dfrac{14w}{-2}$

c $6p \div p$

d $8mn \div 2m$

e $\dfrac{12xy}{3y}$

f $9abc \div bc$

Collecting like terms

You can simplify algebraic expressions by collecting **like terms**.

Like terms contain the same letter, or combination of letters, with the same powers.

a, $4a$ and $-10a$ are all like terms.

b^2, $2b^2$ and $8b^2$ are all like terms.

$2a$, $2ab$ and $2a^2$ are **not** like terms.

ab^2 and a^2b are **not** like terms.

You can add and subtract like terms. To do this you need to consider the $+$ or $-$ sign in front of each term.

NAILIT!

a means '1 lot of a'. You don't need to write $1a$.

DOIT!

Choose two different letters – for example, a and b. On a set of plain cards, write down different terms using those letters, including $+$ and $-$ signs in front of them. Example terms: $3a$, b^2, $2ab$, $-2b$.

Shuffle the cards and deal out 2, 3, 4 or 5 cards so that they form an expression. Simplify each expression where possible.

WORKIT!

Simplify:

a $p + p + p + p$

$4p$

> 4p means '4 lots of p' or '4 × p' or '$p + p + p + p$'.

b $3x + x - 2x$

$2x$

> x means $1x$. So there are $3 + 1 - 2$ lots of x.

When an expression contains two or more different letters, first identify the like terms. This also applies when expressions contain combinations of letters and powers.

WORKIT!

Simplify:

a $4a + 2b + 2a - 4b$

$6a - 2b$

> $4a$ and $2a$ are like terms: $4a + 2a = 6a$.
> $2b$ and $4b$ are like terms: $2b - 4b = -2b$.

b $2f + 3g - f + 4g + 5$

$f + 7g + 5$

> First collect the f terms together and the g terms together. Then deal with the numbers on their own.

You can also simplify expressions that involve surds. If you're unsure about surds have a look at page 34.

$2\sqrt{x} + 5\sqrt{x} = 7\sqrt{x}$ $8\sqrt{x} - 3\sqrt{x} = 5\sqrt{x}$

You can see that $2\sqrt{x}$ means '2 lots of $\sqrt{x}$' or '2 × $\sqrt{x}$' or '$\sqrt{x} + \sqrt{x}$'.

CHECKIT!

1 Simplify

a $f + f + f + f + f$

b $8b - 3b + 2b$

c $4mn + 2mn - mn$

d $4a + 6 - a - 5$

e $3d + 4e + d - 6e$

f $2x + 5y + 3x - 2y - 2$

g $3a - 2b + 4a + 7b$

h $2a - b - 5a - 3$

i $x^2 + x^2$

j $2t^3 + 4 - t^3 - 4$

k $a + a + b \times b$ ← Look carefully at the signs!

l $4\sqrt{x} + 3\sqrt{x}$

m $7\sqrt{x} - 4\sqrt{x}$

n $12\sqrt{x} - \sqrt{x} - 4\sqrt{x}$

Using indices

The terms a^3 and b^4 are written using index notation.

The 3 in a^3 and the 4 in b^4 is called a **power** or **index** (plural: **indices**). It tells you how many times the letter must be multiplied by itself.

a^3 means $a \times a \times a$ and b^4 means $b \times b \times b \times b$.

The a in a^3 and the b in b^4 is called the **base**.

You can simplify algebraic expressions involving powers by using the laws of indices.

You can only use the laws of indices when the bases are the same.

To multiply powers of the same base, add the indices.

WORKIT!

Simplify:

a $x^4 \times x^3$

> Add the indices.
> $x^m \times x^n = x^{m+n}$

$x^4 \times x^3 = x^{4+3} = x^7$

b $2f^2 \times 3f^5$

> Group the numbers and the letters together.
> Multiply the numbers 2 and 3.
> Add the indices.

$2f^2 \times 3f^5 = 2 \times 3 \times f^2 \times f^5 = 6 \times f^{2+5} = 6f^7$

c $4x^2y^3 \times 3xy^2$

$4x^2y^3 \times 3xy^2 = 4 \times 3 \times x^2 \times x \times y^3 \times y^2$

$= 12 \times x^{2+1} \times y^{3+2} = 12x^3y^5$

> Group the numbers (4 and 3) and the same letters (x^2 and x, y^3 and y^2) together.

NAIL IT!

It's always a good idea to show your working.

To divide powers of the same base, subtract the indices.

WORKIT!

Simplify:

a $x^9 \div x$

> If there is no index, the base has a power of 1, so $x = x^1$.

$x^9 \div x = x^{9-1} = x^8$

> Subtract the indices. $x^m \div x^n = x^{m-n}$

b $\dfrac{10p^6}{5p^4}$

$\dfrac{10p^6}{5p^4} = (10 \div 5) \times (p^6 \div p^4) = 2 \times p^{6-4} = 2p^2$

> Group the numbers and the letters together.
> Divide the numbers, then subtract the indices.

c $4x^5 \div 16x^3$

$4x^5 \div 16x^3 = \dfrac{4x^5}{16x^3} = \dfrac{4}{16} \times \dfrac{x^5}{x^3} = \dfrac{1}{4} \times x^{5-3} = \dfrac{x^2}{4}$

> Write the division in fraction form.
> Simplify the numerical fraction: $\frac{4}{16} = \frac{1}{4}$
> Leave the final answer in fraction form (rather than using the decimal form $0.25x^2$).

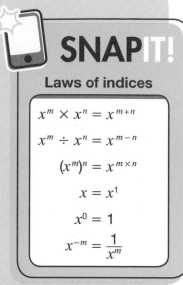

SNAPIT!

Laws of indices

$x^m \times x^n = x^{m+n}$

$x^m \div x^n = x^{m-n}$

$(x^m)^n = x^{m \times n}$

$x = x^1$

$x^0 = 1$

$x^{-m} = \dfrac{1}{x^m}$

To raise a power of a base to a further power, multiply the indices.

WORKIT!

Simplify:

a $(x^4)^3$

$$(x^4)^3 = x^{4 \times 3} = x^{12}$$

> Multiply the indices.
> $(x^m)^n = x^{m \times n}$

b $(2y^3)^2$

$$(2y^3)^2 = 2^2 \times (y^3)^2 = 4 \times y^{3 \times 2} = 4y^6$$

> Both 2 and y^3 are raised to the power of 2, or squared.

> Work out 2^2, then multiply the indices.

Anything raised to the power 0 is equal to 1. So $m^0 = 1$ and $y^0 = 1$.

A negative power means 'one over'.

WORKIT!

Simplify:

a a^{-1}

$$\dfrac{1}{a}$$

> A term raised to a power -1 is 'one over' the term.
> $x^{-1} = \dfrac{1}{x}$

b b^{-3}

$$\dfrac{1}{b^3}$$

> Use $x^{-m} = \dfrac{1}{x^m}$

c $c^{-3} \times c^{-2}$

$$c^{-3} \times c^{-2} = c^{-3 + (-2)} = c^{-5} = \dfrac{1}{c^5}$$

> Add the indices.
> $x^m \times x^n = x^{m+n}$
>
> Be careful with the negative signs:
> $+ -2 = -2$
>
> Give the final answer in simplified form without the negative power.

d $(d^{-3})^{-2}$

$$(d^{-3})^{-2} = d^{-3 \times (-2)} = d^6$$

> Multiply the indices.
> $(x^m)^n = x^{m \times n}$
> negative $\times$ negative = positive

CHECKIT!

1 Simplify:

  a $x^5 \times x^4$

  b $p \times p^4$

  c $2m^4 \times 3m^4$

  d $3m^4n \times 5m^2n^3$

  e $u^{-2} \times u^5$

  f $t^7 \times t^{-6}$

2 Simplify:

  a $x^4 \div x^2$

  b $\dfrac{y^7}{y^3}$

  c $\dfrac{p^9}{p^8}$

  d $8x^6 \div 4x^3$

  e $m^3 \div m^5$

  f $\dfrac{5x^8}{15x^4}$

  g $3x^2 \div 9x$

3 Simplify:

  a $(x^2)^3$

  b $(y^4)^4$

  c $(p^5)^2$

  d $(4m^5)^2$

  e $(x^2)^{-3}$

  f $(n^{-4})^{-2}$

4 Simplify:

  a $4x \times 3x^2$

  b $5x^4 \div x$

  c y^{-2}

  d $a^3b^2 \times a^2b$

5 Simplify fully:

  $\dfrac{x^3 \times x^5}{x^4}$

> First add the indices on the top.

Expanding brackets

Expand single brackets

To **expand** a bracket, multiply **every** term inside the brackets by the term outside the brackets.

$$3 \times x = 3x$$
$$3(x + 2) = 3x + 6$$
$$3 \times 2 = 6$$

Here, there are **two** multiplications to be done when expanding a single bracket and you will obtain **two** terms. If you're unsure about multiplying expressions, go to page 39.

WORKIT!

Expand

a $4(2a + 3)$

$4(2a + 3) = 8a + 12$ ◄──── $4(2a + 3) = (4 \times 2a) + (4 \times 3)$

b $b(b - 4)$

$b(b - 4) = b^2 - 4b$ ◄──── $b(b - 4) = (b \times b) - (b \times 4)$

c $2c(c + 5)$

$2c(c + 5) = 2c^2 + 10c$ ◄──── $2c(c + 5) = (2c \times c) + (2c \times 5)$

To 'expand and simplify', expand the brackets and then collect like terms. Collecting like terms is covered on page 40.

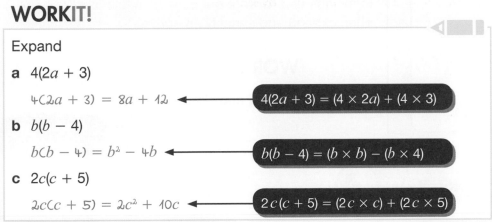

$$2 \times a = 2a \quad 4 \times 2a = 8a$$
$$2(a + 4) + 4(2a - 3) \quad = 2a + 8 + 8a - 12$$
$$= 2a + 8a + 8 - 12$$
$$2 \times 4 = 8 \quad 4 \times -3 = -12 = 10a - 4$$

Each term includes the $+$ or $-$ sign in front of it. So in the first bracket, the number term is $+4$ and in the second bracket it is -3.

You need to take care when there is a $-$ sign outside the brackets.

WORKIT!

Expand and simplify

a $2(3y + 1) - 2(2y - 3)$

$2(3y + 1) - 2(2y - 3)$

$= 6y + 2 - 4y + 6$

$= 2y + 8$

b $3(x + 2y) + 2(x - 3y)$

$= 3x + 6y + 2x - 6y$

$= 5x$

Make sure you answer the question asked. Here you are told to 'expand and simplify' so you need to multiply out the brackets ('expand') and then collect like terms ('simplify').

DO IT!

In your own words, make a note of what is meant by these instructions.

1 Expand

2 Expand and simplify

3 Collect like terms

STRETCH IT!

Expand

a $a(\sqrt{3} + a)$

b $b(\sqrt{5} - b)$

c $(\sqrt{c})^2 + (\sqrt{d})^2$

Remember: '$2\sqrt{x}$' means '2 lots of $\sqrt{x}$' or '$2 \times \sqrt{x}$'.

SNAP IT!

Negative terms outside brackets

When one term is $-$ and one term is $+$ the product is $-$

$$-3 \times p = -3p$$
$$-3(p + 4) = -3p - 12$$
$$-3 \times 4 = -12$$

When one term is $-$ and the other term is $-$ the product is $+$

$$-2 \times q = -2q$$
$$-2(q - 3) = -2q + 6$$
$$-2 \times -3 = 6$$

Expand double brackets

To **expand** two brackets, multiply **every** term inside one bracket by **each** term in the other bracket.

Here there are **four** multiplications to be done when expanding two brackets and you will obtain **four** terms before you collect like terms to simplify the expression.

$x \times 3 = 3x$

$x \times x = x^2$

$(x + 2)(x + 3) = x^2 + 3x + 2x + 6$

$2 \times x = 2x$ $= x^2 + 5x + 6$

$2 \times 3 = 6$

You can use the acronym **FOIL** to help you expand two brackets.

You need to be able to: 'Expand and simplify $(x + 2)^2$'. Here you are being asked to square a single bracket. First you need to write out $(x + 2)^2$ in full: $(x + 2)(x + 2)$. Then expand and simplify as normal.

NAIL IT!

$(a + b)^2$ is **not** equal to $a^2 + b^2$.

SNAP IT! Expanding brackets

To expand two brackets
$(x + 2)(x + 3)$ use:

First term	$(x + 2)(x + 3)$	x^2
Outer term	$(x + 2)(x + 3)$	$+ 3x$
Inner term	$(x + 2)(x + 3)$	$+ 2x$
Last term	$(x + 2)(x + 3)$	$+ 6$

WORKIT!

Expand and simplify $(x - 3)^2$.

$(x - 3)^2 = (x - 3)(x - 3)$ ◄— First write out $(x - 3)^2$ in full.

$= x^2 - 3x - 3x + 9$ ◄—

$= x^2 - 6x + 9$

Use FOIL to find the four terms of the expansion. Then collect like terms as the question is 'expand and simplify'.

Take care with the negative signs:
$x \times -3 = -3x$ $-3 \times -3 = 9$

STRETCH IT!

1 Find the missing terms in this expression.

$(x + 2)(x + \boxed{}) = x^2 + 6x + \boxed{}$

2 Expand and simplify

  a $(x + 3)(2x + 2)$ **d** $(x + 2y)(x - y)$

  b $(3x - 2)(x + 4)$ **e** $(2x - y)(3x + y)$

  c $(2x + 3)(3x - 1)$

✓ CHECK IT!

1 Expand

  a $3(a + 2)$ **f** $-(2y + 4)$

  b $4(b - 4)$ **g** $x(x - 2)$

  c $5(2c + 5)$ **h** $2a(a + 5)$

  d $2(3 - e)$ **i** $3(x + 2y)$

  e $4(x + y + 2)$ **j** $-2(a - b)$

2 Simplify

  a $6a - (3a + 5)$

  b $4x - 6 + 2(x + 5)$

3 Expand and simplify

  a $2(2x + 3) + 4(x + 5)$

  b $3(3y + 1) + 2(4y - 3)$

  c $4(2m + 4) - 3(2m - 5)$

4 Expand and simplify

  a $(x + 2)(x + 3)$ **c** $(a + 3)(a - 7)$

  b $(y - 3)(y + 4)$ **d** $(m - 1)(m - 6)$

5 Expand and simplify

  a $(x + 1)^2$ **c** $(m - 2)^2$

  b $(x - 1)^2$ **d** $(y + 3)^2$

Factorising

Taking out common factors with single brackets

Factorising is the opposite of expanding brackets.
You need to put brackets in.

To factorise an expression	$3a + 12$
1 Find the highest common factor of the terms in the expression.	3 is the highest common factor of the terms $3a$ and 12.
2 Write the factor outside the brackets.	$3a + 12 = 3(\quad)$
3 Work out what you need to multiply the factor by to get each term in the original expression.	$3 \times a = 3a$ and $3 \times 4 = 12$ $3a + 12 = 3(a\quad 4)$
4 Write the + or − sign from the original expression inside the brackets.	$3a + 12 = 3(a + 4)$

Note that 3 is a **factor** of $3a + 12$. Also, $a + 4$ is a factor of $3a + 12$.

Always check your answer by expanding the brackets (see page 43) to see if you get back to the original expression.

$3 \times a = 3a$
$3(a + 4) = 3a + 12$
$3 \times 4 = 12$

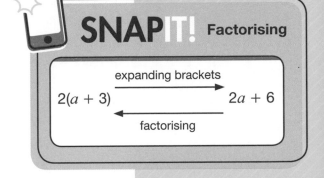

SNAPIT! Factorising

expanding brackets
$2(a + 3) \longrightarrow 2a + 6$
factorising

NAILIT!

The common factor can be a number or a letter or both.

Always check that your answer cannot be factorised further.

WORKIT!

Factorise:

a $2a + 6$

$2a + 6 = 2(a + 3)$

Take 2 outside the bracket. It is the highest common factor of 2 and 6.

b $8b - 12$

$8b - 12 = 4(2b - 3)$

The common factors of 8 and 12 are 2 and 4. The highest common factor is 4.

$2(4b - 6)$ is **not** fully factorised.

c $5xy + 15x$

$5xy + 15x = 5x(y + 3)$

The common factors of $5xy$ and $15x$ are x and $5x$. The highest common factor is $5x$.

$x(5y + 15)$ is **not** fully factorised.

d $c^2 + 8c$

$c^2 + 8c = c(c + 8)$

c is the common factor of c^2 ($c \times c$) and $8c$ ($8 \times c$).

e $ad + bd$

$ad + bd = d(a + b)$

d is the common factor of ad and bd.

Factorising quadratic expressions

A **quadratic expression** has a squared term (x^2) and no higher power.

The following are examples of quadratic expressions of the form $x^2 + bx + c$ where b and c are numbers.

$$x^2 + 4x + 4 \qquad x^2 + 3x - 18 \qquad x^2 - 6x + 5 \qquad x^2 - 2x - 15$$

To factorise a quadratic expression of this type you need to write the expression with two brackets: $(x \quad)(x \quad)$. An x in each bracket will give x^2 when multiplied.

To factorise the quadratic expression $x^2 + 5x + 6$ you need to find two numbers that multiply to make +6 and add up to +5:

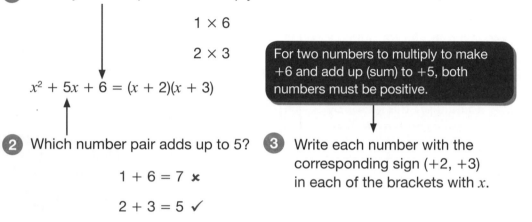

1 Identify number pairs that multiply to make 6:

$$1 \times 6$$

$$2 \times 3$$

> For two numbers to multiply to make +6 and add up (sum) to +5, both numbers must be positive.

$$x^2 + 5x + 6 = (x + 2)(x + 3)$$

2 Which number pair adds up to 5?

$$1 + 6 = 7 \text{ ✗}$$

$$2 + 3 = 5 \text{ ✓}$$

3 Write each number with the corresponding sign (+2, +3) in each of the brackets with x.

4 Check your answer by expanding the brackets to see if you get back to the original expression.

$$(x + 2)(x + 3) = x^2 + 3x + 2x + 6$$

$$= x^2 + 5x + 6$$

> If you're unsure about expanding double brackets, have a look at page 44.

WORKIT!

Factorise $x^2 + 2x - 3$

$x^2 + 2x - 3$

Number pairs that multiply to make -3:

$1 \times -3, -1 \times 3$

Sum of number pairs:

$1 + -3 = -2$ ✗

$-1 + 3 = 2$ ✓

$x^2 + 2x - 3 = (x - 1)(x + 3)$

> For two numbers to multiply to make -3, one number must be negative and one must be positive.

WORKIT!

Factorise $x^2 - 6x + 5$

$x^2 - 6x + 5$

Number pairs that multiply to make 5:

$1 \times 5, -1 \times -5$

Sum of number pairs:

$1 + 5 = 6$ ✗

$-1 + -5 = -6$ ✓

$x^2 - 6x + 5 = (x - 1)(x - 5)$ ◀

For two numbers to multiply to make +5 and add up to −6, both numbers must be negative.

SNAPIT!   Difference of two squares

Factorise quadratic expressions of the form $x^2 - a^2$ using the rule:

$x^2 - a^2 \equiv (x + a)(x - a)$

WORKIT!

Factorise $x^2 - 25$

$x^2 - 25 = x^2 - 5^2$ ◀

$= (x + 5)(x - 5)$

25 is a square number: $25 = 5^2$
Use $x^2 - a^2 = (x + a)(x - a)$ with $a = 5$

STRETCHIT!

The area of a rectangle is $x^2 + 3x + 2$.

The length of the rectangle is $x + 2$.

Find an expression for the width of the rectangle.

STRETCHIT!

Factorise:

a $a^2 - 3$ ◀

b $b^2 - 5$

$a^2 - 3 = a^2 - (\sqrt{3})^2$ using the fact that $\sqrt{3} \times \sqrt{3} = 3$

✓ CHECKIT!

1 Factorise

  a $3a + 9$ **c** $7 + 14c$

  b $5b - 10$ **d** $d^2 - 2d$

2 Factorise

  a $8a + 20$ **c** $18 + 9c$

  b $4b - 12$ **d** $2d^2 - 3d$

3 Factorise

  a $4x - 6y$ **e** $2n - 9n^2$

  b $am + bm$ **f** $5x + 10xy$

  c $12a^2 + 8a$ **g** $4pq - 12p$

  d $4x^2 + 3xy$ **h** $4x^2y - 8y$

4 The expression $4(x - 3) + 3(2x + 6)$ simplifies to $a(5x + b)$.

  Work out the values of a and b.

5 Factorise

  a $x^2 + 8x + 7$ **e** $x^2 - 6x + 9$

  b $x^2 + 4x - 5$ **f** $x^2 + 7x + 12$

  c $x^2 - 2x - 8$ **g** $x^2 + 3x - 10$

  d $x^2 - 5x + 6$ **h** $x^2 - x - 20$

6 Factorise

  a $x^2 - 16$ **c** $x^2 - 81$

  b $x^2 - 36$ **d** $y^2 - 100$

Substituting into expressions

SNAPIT! BIDMAS

Use **BIDMAS** to remember the correct order of operations.

Brackets
Indices
Division
Multiplication
Addition
▼ Subtraction

$x + y$, $2a + 3$, $4mn$, $3x^2 - 2$, $3(p + 1)$ and $5pq + pqr$ are all examples of algebraic **expressions**. The letters are used to represent unknown numbers.

If you are given the value of each of the letters, you can **substitute** them into the expression to work out the value of the expression.

You need to use the correct order of operations when doing the calculation. If you're unsure about this, see page 31.

The value of the letters may be positive or negative.

Use brackets when substituting a negative number.
$5p$ means $5 \times p$.
If $p = 2$ then
$5p = 5 \times 2 = 10$
Work out each multiplication first (BID**M**AS).

WORKIT!

1 Work out the value of $5p - 2q$ when $p = 2$ and $q = -3$

$5 \times 2 - 2 \times (-3) = 10 - (-6)$

$= 10 + 6$ ← Subtracting a negative number is the same as adding.

$= 16$

2 Work out the value of $r(s - 4)$ when $r = 5$ and $s = -4$

$5 \times (-4 - 4) = 5 \times -8$

$= -40$ ← Work out the brackets first (**B**IDMAS).
'positive' × 'negative' = 'negative'

Take care when substituting negative numbers into expressions involving powers.

NAILIT!

Substitute all the values **before** doing any calculations.

WORKIT!

1 Work out the value of $m^2 + n^2$ when $m = 3$ and $n = -1$

$3^2 + (-1)^2 = 9 + 1$ ← Work out each power (indices) first (BIDMAS).
$(-1)^2 = -1 \times -1 = 1$
'negative' × 'negative' = 'positive'

$= 10$

2 Work out the value of $5x^3$ when $x = -2$

Only the value of x is cubed: $5x^3$ is not the same as $(5x)^3$.
$(-2)^3 = -2 \times -2 \times -2$
$= -8$
'negative' × 'negative' × 'negative' = 'negative'

$5 \times (-2)^3 = 5 \times -8$

$= -40$

You will also need to be able to substitute decimals and fractions. Care should be taken when the expression is in fraction form.

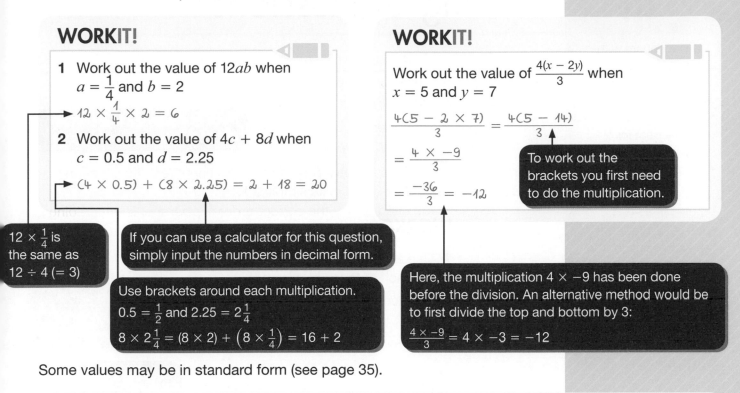

WORKIT!

1 Work out the value of $12ab$ when
   $a = \frac{1}{4}$ and $b = 2$

   $12 \times \frac{1}{4} \times 2 = 6$

2 Work out the value of $4c + 8d$ when
   $c = 0.5$ and $d = 2.25$

   $(4 \times 0.5) + (8 \times 2.25) = 2 + 18 = 20$

$12 \times \frac{1}{4}$ is the same as $12 \div 4 (= 3)$

If you can use a calculator for this question, simply input the numbers in decimal form.

Use brackets around each multiplication.
$0.5 = \frac{1}{2}$ and $2.25 = 2\frac{1}{4}$
$8 \times 2\frac{1}{4} = (8 \times 2) + \left(8 \times \frac{1}{4}\right) = 16 + 2$

WORKIT!

Work out the value of $\frac{4(x - 2y)}{3}$ when
$x = 5$ and $y = 7$

$\frac{4(5 - 2 \times 7)}{3} = \frac{4(5 - 14)}{3}$

$= \frac{4 \times -9}{3}$

$= \frac{-36}{3} = -12$

To work out the brackets you first need to do the multiplication.

Here, the multiplication 4×-9 has been done before the division. An alternative method would be to first divide the top and bottom by 3:
$\frac{4 \times -9}{3} = 4 \times -3 = -12$

Some values may be in standard form (see page 35).

WORKIT!

1 Work out the value of $5e$ when $e = 4 \times 10^3$

   $5 \times 4 \times 10 \times 10 \times 10 = 20\,000$

2 Work out the value of $8f + 10g$
   when $f = 5 \times 10^{-2}$ and $g = 6 \times 10^{-2}$

   $8 \times 5 \div (10 \times 10) + 10 \times 6 \div (10 \times 10)$

   $= (40 \div 100) + (60 \div 100) = 0.4 + 0.6 = 1$

$10^3 = 10 \times 10 \times 10$

In standard form, $20\,000 = 2 \times 10^4$

$10^{-2} = \frac{1}{10 \times 10} = \frac{1}{100}$

$8 \times 5 \times 10^{-2}$ could be written as $\frac{8 \times 5}{100} = \frac{40}{100}$

✓ CHECKIT!

1 Work out the value of $5a + 2b$ when
   $a = 3$ and $b = -2$.

2 Work out the value of each expression
   when $x = 2$ and $y = -4$.

   a $x - 2y$ d $x^2 + y^2$

   b $3xy$ e $2x + 4(x - y)$

   c $4y - 3x$ f $\frac{1}{2}(x + y)$

3 True or false: $3a^2 = 81$ when $a = 3$?

   Give a reason for your answer.

4 Work out the value of each expression
   when $p = \frac{1}{2}$ and $q = -4$.

   a $10pq$ c $\frac{q}{p}$

   b $8p^2$
                      d $2q^2 - 12p$

5 Work out the value of $\frac{d(e - 2)}{f}$
   when $d = 7$, $e = -3$ and $f = 10$.

Writing expressions

DO IT!

1 Draw a square with sides of length x. Write down an expression for its perimeter and area. Repeat for a rectangle with length 2 and width $x + 2$.

2 Draw a triangle with angles a, $a - 20$, $a + 50$. Write down an expression for the sum of its angles.

For an unknown number x, the expression $x + 3$ means '3 more than x' or 'x with 3 added' or 'add 3 to x'.

To write expressions correctly you need to use the notation and symbols with care.

> If you're unsure about algebraic notation, have a look at page 39.

When writing expressions to solve problems or represent real-life situations, the expression should be given in its simplest form.

STRETCH IT!

The perimeter of the triangle is equal to the perimeter of the square.

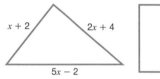

$x + 2$ $2x + 4$ $5x - 2$

a Write down an expression for the perimeter of the triangle. Give your answer in its simplest form.

b Find an expression for the length of the side of the square in terms of x.

> First factorise your expression from part a.

WORKIT!

1 Use algebra to write down an expression for each of these numbers.

a 4 less than a

$a - 4$

> $a - 4$ means '4 less than a'. $a - 4$ is **not** the same as $4 - a$.

b a multiplied by 2 and then 4 added

$2a + 4$

c a plus 4 and then multiplied by 2

$2(a + 4)$

> Use a bracket to show that '$a + 4$' is multiplied by 2.
>
> $2a + 4$ and $2(a + 4)$ are **not** the same.

WORKIT!

The width of a rectangle is x cm. Its length is 3 times its width. Write down an expression for its perimeter.

Length $= 3x$

> First write down an expression for the length. $3x$ means '3 times x'.

Perimeter $= x + 3x + x + 3x$

$= 8x$ cm

> Simplify the expression by collecting like terms (see page 40).

> Perimeter is the distance around the shape. You are adding 2 widths and 2 lengths so you could also write the expression as:
> $2 \times x + 2 \times 3x = 2x + 6x = 8x$

CHECK IT!

1 Use algebra to write down an expression for each of these numbers.

a q less than 4   c x multiplied by y

b m more than n d p multiplied by itself

2 Nathan buys x white T-shirts and y patterned T-shirts for his holiday. Write down an expression for the total number of T-shirts he buys.

3 £y is shared equally between eight people. How much does each person receive?

4 Write an expression for the total cost, in pounds, of n necklaces at £100 each and b bracelets at £75 each.

5 A triangle has sides of length $3a$, $2a + 4$ and $4a - 2$ cm.

Write down an expression for its perimeter. Give your answer in its simplest form.

6 Here is a right-angled triangle.

$2x + 5$

4

Write down an expression for its area.

Solving linear equations

Solving simple equations

$2x + 2 = 8$ is a **linear equation**. An equation is only true for certain values of the letter. You can **solve** the equation to find the value of the letter.

To solve an equation you need to rearrange it so that the letter is on its own on one side of the equation. To rearrange an equation you can:

- add the same number to both sides

  $a - 4 = 12$

  $a = 12 + 4$ ← Add 4 to both sides.

  $a = 16$

- subtract the same number from both sides

  $b + 8 = 13$

  $b = 13 - 8$ ← Subtract 8 from both sides.

  $b = 5$

- divide both sides by the same number

  $4c = 20$ ← $4c$ means $4 \times c$

  $c = \dfrac{20}{4}$ ← Divide both sides by 4.

  $c = 5$

- multiply both sides by the same number

  $\dfrac{d}{8} = 3$ ← $\dfrac{d}{8}$ means $d \div 8$

  $d = 3 \times 8$ ← Multiply both sides by 8.

  $d = 24$

NAILIT!

The solution to an equation can be a positive number, negative number, fraction or decimal.

When solving an equation with two operations, look to do the + or − first.

> You need to do two operations to get x on its own. Do them one at a time, dealing with the '− 2' first.

WORKIT!

Solve

a $4x - 2 = 8$

  $4x = 10$ ← Add 2 to both sides.

  $x = 2.5$ ← Divide both sides by 4. The answer can be given as a decimal ($x = 2.5$) or as a mixed number $\left(x = 2\frac{1}{2}\right)$.

b $\dfrac{p}{6} + 2 = 7$

  $\dfrac{p}{6} = 5$ ← Subtract 2 from both sides.

  $p = 30$ ← Multiply both sides by 6.

c $\dfrac{x + 5}{2} = 6$

  $x + 5 = 12$ ← Multiply both sides by 2.

  $x = 7$ ← Subtract 5 from both sides.

Take care when the **coefficient** (the number in front of the letter) is negative.

> The first step is to make the coefficient of x positive.

> Subtract 5 from both sides.

WORKIT!

Solve $13 - 4x = 5$

  $13 - 4x = 5$

  $13 = 5 + 4x$ ← Add $4x$ to both sides.

  $8 = 4x$

  $2 = x$ ← Divide both sides by 4. You can write the answer as '$2 = x$' or '$x = 2$'.

DOIT!

Solve the Work it! equation $13 - 4x = 5$ using these alternative steps:

Step 1: subtract 13 from both sides.

Step 2: divide both sides by -4.

Solving equations with brackets and the unknown on both sides

To solve an equation with brackets, expand the brackets first. If you're unsure about expanding brackets, have a look at page 43.

$2(x + 3) = 7$

$2x + 6 = 7$ ← $2(x + 3) = 2 \times x + 2 \times 3$

$2x = 1$ ← Subtract 6 from both sides.

$x = \frac{1}{2}$ ← Divide both sides by 2.

NAIL IT!

When solving equations:
- show all your working
- use a new line for each step of working
- do the same thing to **both** sides of the equation.

WORKIT!

Solve $2(3m + 1) + 6 = 2$

$2(3m + 1) + 6 = 2$

$6m + 2 + 6 = 2$ ← Expand brackets.

$6m + 8 = 2$ ← Collect like terms.

$6m = -6$ ← Subtract 8 from both sides.

$m = -1$ ← Divide both sides by 6.

In an equation where the letter appears on both sides, the first step is to rearrange the equation so that the letter appears on **one** side only.

> Both x terms ($3x$ and x) have a positive coefficient. Subtract the smaller term in x from both sides so that the coefficient of x remains positive.

$3x + 9 = x + 5$

$2x + 9 = 5$

$2x = -4$ ← Subtract 9 from both sides.

$x = -2$ ← Divide both sides by 2.

NAIL IT!

When solving linear equations with the unknown on both sides, collect the terms in the unknown on the side of the equation that gives them a positive coefficient.

WORKIT!

Solve $5 - 3x = 7 - 4x$

$5 - 3x = 7 - 4x$ ← Both x terms ($-3x$ and $-4x$) have a negative coefficient. Add the higher term in x ($4x$) to both sides so that the coefficient of x becomes positive.

$5 + x = 7$

$x = 2$

Subtract 5 from both sides.

WORKIT!

Solve $4(2x + 1) = 3(5x - 1)$

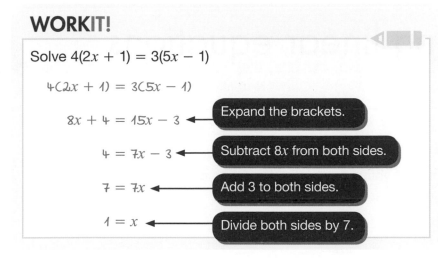

$4(2x + 1) = 3(5x - 1)$

$8x + 4 = 15x - 3$ ← Expand the brackets.

$4 = 7x - 3$ ← Subtract $8x$ from both sides.

$7 = 7x$ ← Add 3 to both sides.

$1 = x$ ← Divide both sides by 7.

DOIT!

Create a mind map for 'Solving linear equations'. Include an example of all the different equation types and illustrate a worked example for each type with all the key points.

CHECKIT!

1 Solve

  a $5a = 35$

  b $b - 9 = 8$

  c $\frac{c}{4} = 4$

  d $d + 4 = 2$

2 Solve

  a $2x + 3 = 13$

  b $3y - 4 = 11$

  c $2p + 9 = 1$

  d $\frac{f}{3} - 7 = 4$

  e $\frac{x + 5}{2} = 8$

  f $\frac{f - 7}{3} = 4$

3 Solve

  a $9 - m = 7$

  b $10 - 3x = 1$

  c $7 - 2x = 2$

  d $5 = 1 - 2f$

4 Hannah solves the equation $2x + 4 = 8$

Here is her working:

$2x + 4 = 8$

$2x = 8$

$x = 4$

Hannah's answer is wrong.
What mistake has she made?

5 Solve

  a $3(a + 2) = 15$

  b $4(b - 2) = 4$

  c $3(4c - 9) = 9$

  d $2(d + 3) + 4 = 2$

  e $4(2x + 3) - 2 = 6$

6 Solve

  a $3m = m + 6$

  b $5t - 6 = 2t + 3$

  c $4x + 3 = 2x + 8$

  d $3 - 2p = 6 - 3p$

  e $3y - 8 = 5y + 4$

7 Solve

  a $2(x + 5) = x + 6$

  b $5(a - 1) = 4 - a$

  c $7b - 2 = 2(b + 4)$

  d $4(2y + 1) = 3(5y - 1)$

  e $2x - 1 = 8 - 4x$

Writing linear equations

Using algebra to solve angle or perimeter problems

For some problems you will use the information given in the question to set up an equation.

WORKIT!

The diagram shows a quadrilateral.

All measurements are in centimetres.

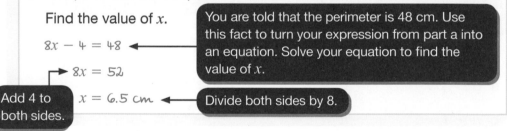

a Write down an expression, in terms of x, for the perimeter of the quadrilateral.

$$\text{Perimeter} = 3x + 2x + 3 + x + 2x - 7$$
$$= 8x - 4$$

Simplify the expression by collecting like terms (page 40). x means '$1x$'.

b The perimeter of the quadrilateral is 48 cm.

Find the value of x.

$$8x - 4 = 48$$
$$8x = 52$$
$$x = 6.5 \text{ cm}$$

You are told that the perimeter is 48 cm. Use this fact to turn your expression from part a into an equation. Solve your equation to find the value of x.

Add 4 to both sides.

Divide both sides by 8.

For some problems you will need to use your knowledge of angles and properties of shapes to set up an equation.

WORKIT!

Here is a triangle.

Work out the size of the smallest angle.

Write an equation in terms of a using the fact that the angle sum of a triangle is 180°.

$$a + 20 + a + 25 + a - 15 = 180$$

$$3a + 30 = 180$$

Subtract 30 from both sides.

$$3a = 150$$

Divide both sides by 3.

$$a = 50$$

$$a - 15 = 50 - 15 = 35$$

The smallest angle is 35°.

STRETCHIT!

Here is a square.

3(x + 3) cm

5x − 12 cm

Work out the length of the side of the square.

Make sure you answer the question asked. The smallest angle is given by the expression $a - 15$.

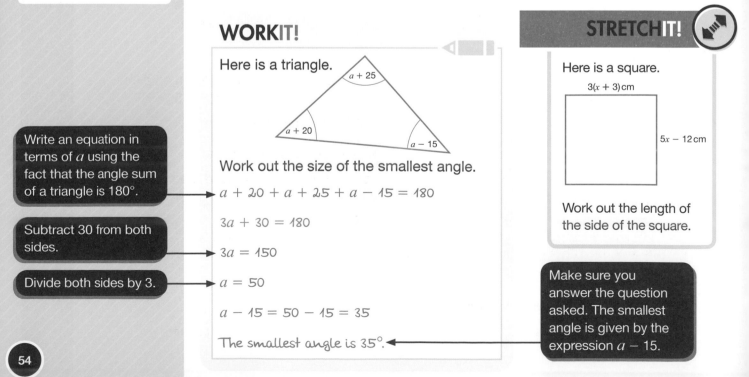

Using algebra to solve word problems

For some word problems you need to choose your own letter to represent the unknown. State the letter you have chosen and what it represents at the start of your working. For example:

'Imelda has 3 more sweets than Jack. Katerina has twice as many sweets as Imelda. They have 25 sweets in total. Work out how many sweets Imelda has.'

You can use any letter to represent the unknown. → Let n = number of sweets that Jack has.

$$n + n + 3 + 2(n + 3) = 25$$

Form an equation for the total number of sweets (which equals 25).
Jack has n sweets.
Imelda has $n + 3$ ('3 more' than Jack)
Katerina has $2(n + 3)$ ('twice as many' as Imelda)

Expand the brackets. → $n + n + 3 + 2n + 6 = 25$

Collect like terms. → $4n + 9 = 25$

Subtract 9 from both sides. → $4n = 16$

Divide both sides by 4. → $n = 4$

Substitute $n = 4$ to find the number of sweets Imelda has.

Jack has 4 sweets.

→ Imelda has $n + 3 = 4 + 3 = 7$ sweets.

✓ CHECKIT!

1 A square has sides of length $2s + 3$ cm.

  a Write down an expression, in terms of s, for the perimeter of the square.

  b The perimeter of the square is 84 cm. Find the value of s.

2 Here is a quadrilateral.

  a Work out the value of x.

  b Work out the size of the largest angle.

3 Monica is 4 years younger than Karen.

  The sum of their ages is 64.

  Find Monica's age.

4 Multiplying a number by 2 and then adding 4 gives the same answer as subtracting the number from 16.

  Find the number.

5 The width of a rectangle is 2 cm smaller than its length. Its perimeter is 36 cm.

  Find the width of the rectangle.

6 Here is an isosceles triangle.

  Work out the value of a and the value of b.

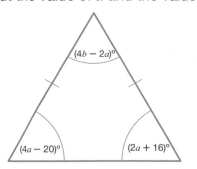

Linear inequalities

The wide end (or 'mouth') of the inequality always points towards the larger value.

Picture a number line: −5 is to the left of −3.

Inequalities and number lines

Inequalities can be used to compare quantities. They are written using inequality signs.

$4 > 2$ is a true statement as 4 is greater than 2.

$−5 < −3$ is a true statement as −5 is less than −3.

For $a < b$ to be a true statement, the value of a would have to be less than the value of b.

For $x \geq 2$ to be a true statement, the value of x would have to be greater than or equal to 2. So if x is an integer it could be 2 or 3 or 4 or 5 and so on.

SNAPIT! Inequality signs

> means 'is greater than'
≥ means 'is greater than or equal to'
< means 'is less than'
≤ means 'is less than or equal to'

DOIT!

Choose a positive or negative integer for a. Choose a positive or negative integer for b. Draw number lines to show these inequalities. Label each inequality on your diagram.

1 $x > a, x < b$
2 $x \geq a, x \leq b$
3 $a < x < b$
4 $a < x \leq b$
5 $a \leq x < b$
6 $a \leq x \leq b$

An **integer** is a positive or negative whole number, including 0.

WORKIT!

Write down all the integer values of x that satisfy the inequality $−3 \leq x < 2$

$x = −3, −2, −1, 0, 1$

$−3 \leq x < 2$ means x is greater than or equal to −3 but less than 2. So −3 is included but 2 isn't.

An inequality can be shown on a number line.

$x > −2$

The empty (or open) circle shows that −2 is **not** included.

$x \leq 2$

The filled (or closed) circle shows that 2 **is** included.

WORKIT!

a Write down the inequality represented on the number line.

$x \geq −4$ The circle is filled so −4 **is** included.

b Show the inequality $−2 < x \leq 3$ on a number line.

$−2 < x \leq 3$ means x is greater than −2 but less than or equal to 3. All the values between −2 and 3 must be shown on the number line.

−2 is not included so use an empty circle.

3 is included so use a filled circle.

Solving inequalities

You can solve inequalities in the same way that you solve linear equations.

Use the inequality sign on each line of your working.

You may need to represent the solution set on a number line.

If you're unsure about solving equations, have a look at page 51.

WORKIT!

a Solve the inequality $2x + 6 \geq 1$

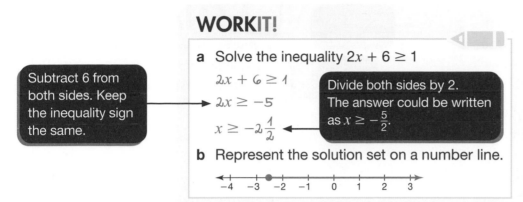

Subtract 6 from both sides. Keep the inequality sign the same.

$2x + 6 \geq 1$

$2x \geq -5$

$x \geq -2\frac{1}{2}$

Divide both sides by 2. The answer could be written as $x \geq -\frac{5}{2}$.

b Represent the solution set on a number line.

You also need to be able to solve inequalities where the unknown appears on both sides of the inequality sign.

WORKIT!

a Solve the inequality $3x - 4 < 5x - 6$

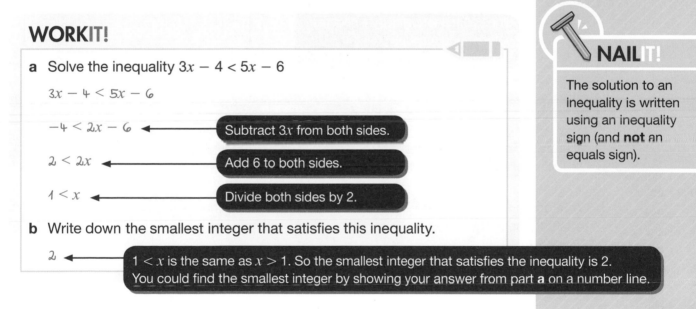

$3x - 4 < 5x - 6$

$-4 < 2x - 6$ ← Subtract $3x$ from both sides.

$2 < 2x$ ← Add 6 to both sides.

$1 < x$ ← Divide both sides by 2.

b Write down the smallest integer that satisfies this inequality.

2 ← $1 < x$ is the same as $x > 1$. So the smallest integer that satisfies the inequality is 2. You could find the smallest integer by showing your answer from part **a** on a number line.

NAILIT!

The solution to an inequality is written using an inequality sign (and **not** an equals sign).

You also need to be able to solve two inequalities in x.

WORKIT!

a Find all the integer values of x that satisfy the inequality
$-3 < 2x + 1 < 5$

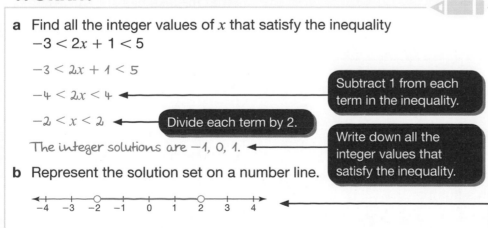

$-3 < 2x + 1 < 5$

$-4 < 2x < 4$ ← Subtract 1 from each term in the inequality.

$-2 < x < 2$ ← Divide each term by 2.

The integer solutions are $-1, 0, 1$. ← Write down all the integer values that satisfy the inequality.

b Represent the solution set on a number line.

You could use your number line to write down the integer solutions for part a. Or use it to check your answer to part a!

Care must be taken when there is a negative sign in front of the unknown.

To solve $8 - 3x > 2$ you can choose between two methods.

The first method involves adding $3x$ to both sides to ensure the coefficient of x is not negative.

$8 - 3x > 2$

$8 > 2 + 3x$ ◄ **Add $3x$ to both sides.**

Subtract 2 from both sides. ► $6 > 3x$

$2 > x$ or $x < 2$ ◄ **Divide both sides by 3.**

The second method involves a division by -3.

$8 - 3x > 2$

Subtract 8 from both sides. ► $-3x > -6$

Divide both sides by -3. ► $x < 2$

When you multiply or divide both sides of an inequality by a negative number you must reverse the inequality sign.

SNAPIT!

Dividing through by -1 means you **reverse the inequality sign**.

$-x > -2$

Everything in the first diagram is reflected to get the second one, including the direction of the arrow.

$x < 2$

CHECKIT!

1 Write down all the integer values of x that satisfy these inequalities.

a $2 < x < 6$ c $-1 < x \le 3$

b $2 \le x < 6$ d $-3 \le x \le 1$

2 Write down the inequality represented on each number line.

a

b

c

3 Show each inequality on a number line.

a $x \ge 1$ c $-2 \le x < 2$

b $1 < x \le 4$ d $-3 \le x \le 0$

4 Solve these inequalities. Represent each solution set on a number line.

a $2x - 2 > 4$ c $4x < 2x - 10$

b $4x + 3 \le 13$ d $7x + 2 \ge 3x - 2$

5 Olivia has solved the inequality $3(x + 4) > 22$.
Here is her working:

$3(x + 4) > 22$

$3x + 4 > 22$

$3x > 18$

$x > 6$

Olivia's answer is wrong.

What mistake has she made?

6 Write down all the integers that satisfy these inequalities.

a $-12 < 4x \le 8$ c $-6 < 6x \le 18$

b $-8 \le 2x < 14$ d $9 \le 3n \le 15$

7 a Solve the inequality $2x + 3 < 4x + 8$.

b Write down the smallest integer value of x that satisfies the inequality.

8 Solve these inequalities.

a $4 - x \le 1$ c $8 - 2x \ge 7$

b $6 - 3x > 9$ d $-2 < -x \le 3$

Formulae

Using formulae

A **formula** (plural: **formulae**) is a mathematical rule. It gives the relationship between different quantities.

The word formula for the area of this rectangle is:

area = length × width

The same formula can be written using algebra as: $A = lw$

You can use a formula to calculate one quantity if you know the other quantities. You can substitute the values of length (l) and width (w) into this formula to find the area (A).

width

length

DOIT!

Make a poster to show all the formulae that you need to know for your maths exam.

WORKIT!

Jazz uses this rule to work out her mobile phone bill:

Total bill = monthly charge + cost per minute of calls × number of minutes of calls.

The monthly charge is £12. For calls in excess of her contract limit, the cost per minute is 7p.

Jazz made 80 minutes of calls above her contract limit.

Work out her total bill.

7p = £0.07 ◀── The values substituted into a formula must be in the same units. Write 7p in pounds.

Total bill = £12 + £0.07 × 80 ◀──

  = £12 + £5.60

  = £17.60

Substitute all the values before doing any calculations. Use the correct order of operations: **M**ultiplication before **A**ddition (BID**MA**S).

WORKIT!

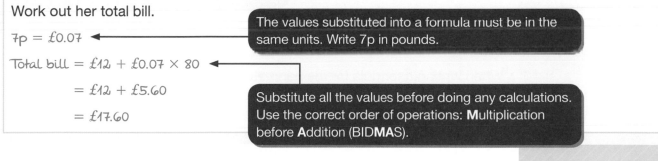

This formula can be used to work out the distance, s, moved by an object.

$s = ut + \frac{1}{2}at^2$

The velocity of an object is its speed in a particular direction.

where u is velocity, t is time and a is acceleration.

Find s when $u = 5$, $t = 4$ and $a = 10$.

$s = 5 \times 4 + \frac{1}{2} \times 10 \times 4^2$ ◀── Substitute the values of u, t and a into the formula.

$= 5 \times 4 + \frac{1}{2} \times 10 \times 16$ ◀── Work out the power (or indices) first (BIDMAS). Then work out the multiplications (BIDMAS). Do the addition last (BIDMAS).

$= 20 + 80$

$= 100$

Use a new line for each step of working.

NAILIT!

To find a quantity using a formula:

• use the same units of measurement

• use the correct order of operations: BIDMAS (see page 31).

Writing formulae

You can use information given in words to write a word formula or an algebraic formula. This can then be used to solve a problem.

WORKIT!

This rule can be used to work out the cooking time of a turkey:

Cooking time in minutes = 15 × weight in kg + 70

A turkey weighs w kg and is cooked for T minutes.

Write down a formula for T in terms of w.

$15w$ means '15 × w'.

$T = 15w + 70$ ← You do not need to include the units in the algebraic formula.

WORKIT!

The cost of printing promotional leaflets is £150 plus 2p per leaflet.

The cost of printing n leaflets is C pounds.

Write down a formula for C in terms of n.

$2p = £0.02$ ← The cost is in pounds so write 2p in pounds.

$C = 150 + 0.02n$

When writing a formula for the perimeter or area of a shape you should give the formula in its simplest form. To do this, you need to be able to collect like terms (page 40) and multiply expressions (page 39).

DOIT!

A regular polygon has sides of equal length, l. Write down a formula for the perimeter, P, of a regular polygon with five sides.

The perimeter is the distance around a shape.

WORKIT!

The diagram shows an isosceles triangle.

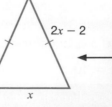

$2x - 2$

An isosceles triangle has two equal sides (as shown by the dashes on two of the sides).

x

Write down a formula for the perimeter P of the triangle in terms of x.

Write down all three sides before collecting like terms to simplify the formula.

$P = 2x - 2 + 2x - 2 + x$

$P = 5x - 4$

Rearranging formulae

The letter on its own on one side of the equals sign in a formula is called the **subject**. In the formula $A = lw$ the subject is A.

To find the value of a quantity that is not the subject of the formula, substitute the given values into the formula and then solve the equation.

If you're unsure about solving equations, see page 51.

WORKIT!

This formula can be used to work out the amount Tomasz the builder is paid for a job:

$P = 80 + 20h$

where P is the amount paid in pounds and h is the number of hours worked.

Tomasz is paid £520 for a job.

How many hours did he work?

$520 = 80 + 20h$

$440 = 20h$ ← Subtract 80 from both sides.

$22 = h$ ← Divide both sides by 20.

Tomasz worked 22 hours.

You can check your answer by substituting the value of h back into the formula:

$P = 80 + 20 \times 22$

$= 80 + 440$

$= 520 ✓$

To change the subject of a formula you have to **do the same thing to both sides** of the formula until you have the required letter on its own on one side.

WORKIT!

Rearrange each formula to make the letter in square brackets the subject.

a $y = mx + c$ $[m]$ ← You need to get m on its own on the right-hand side (RHS).

$y = mx + c$ ← First subtract the terms you don't need on the RHS $(-c)$.

$y - c = mx$ ← mx means '$m \times x$' so divide by x to get m on its own.

$\dfrac{y - c}{x} = m$ ← Everything on the left-hand side (LHS) is divided by x.

b $A = \dfrac{1}{2}bh$ $[b]$

$A = \dfrac{1}{2}bh$

$2A = bh$ ← Multiply both sides by 2 to get rid of the fraction.

$\dfrac{2A}{h} = b$ ← Divide both sides by h.

c $a = \dfrac{b}{c} + d$ $[b]$

$a = \dfrac{b}{c} + d$ ← Deal with the '$+ d$' term first.

$a - d = \dfrac{b}{c}$

$c(a - d) = b$ ← Everything on the LHS is multiplied by c. You need to use brackets to show this.

d $A = \pi r^2$ $[r]$

$A = \pi r^2$

$\dfrac{A}{\pi} = r^2$ ← Divide both sides by π.

$\sqrt{\dfrac{A}{\pi}} = r$ ← Square root both sides to get r as the subject.

e $P = 2(l + w)$ $[l]$

$P = 2(l + w)$

$P = 2l + 2w$ ← Expand the brackets first.

$P - 2w = 2l$ ← Subtract $2w$ from both sides.

$\dfrac{P - 2w}{2} = l$ ← Divide both sides by 2.

Or $\dfrac{P}{2} - w = l$

f $d = \sqrt{2Rh}$ $[h]$

$d = \sqrt{2Rh}$

$d^2 = 2Rh$ ← Square both sides using the result that $\sqrt{x} \times \sqrt{x} = x$

$\dfrac{d^2}{2R} = h$ ← Divide both sides by $2R$.

In some cases, the subject appears twice in the formula.

WORKIT!

Make x the subject of the formula $4xy = x + 1$

$4xy = x + 1$ ← First, you need to get all the terms that contain the subject x together on one side.

$4xy - x = 1$

Subtract x from both sides.

$x(4y - 1) = 1$ ← Factorise the left-hand side. (See page 45 for a reminder about factorising.)

$x = \dfrac{1}{4y - 1}$ ← Divide both sides by $4y - 1$.

STRETCHIT!

Show that $4a = 5(2b^2 - a)$ can be rearranged to $b = \sqrt{\dfrac{9a}{10}}$.

DOIT!

Make l the subject of the formula
$P = 2(l + w)$ using this method.

1 Divide both sides by 2.

2 Subtract w from both sides.

Compare this method with that shown in the top Work it! part e.

 **CHECKIT!**

1 Sadik uses this rule to work out his pay:

Pay = rate per hour × number of hours worked + commission

Sadik's rate per hour is £8. He worked for 35 hours and his commission was £25.

Work out his pay.

2 This formula can be used to work out the perimeter of a rectangle:

$P = 2(l + w)$

Work out the value of P when $l = 8$ and $w = 5.5$.

3 Convert 45°C to °F using the formula

$F = \frac{9}{5} C + 32$

4 This formula can be used to work out the final velocity, v, of an object:

$v = u + at$ where u is initial velocity, a is acceleration and t is time.

Work out the value of v when $u = 10$, $a = -20$ and $t = 5$.

5 This formula can be used to work out the final velocity, v, of an object:

$v^2 = u^2 + 2as$ where u is initial velocity, a is acceleration and s is displacement.

Work out the value of v when $u = 2.5$, $a = -9.8$ and $s = 0.2$.

Give your answer correct to 1 decimal place.

Take care – you are asked to find v not v^2.

6 This rule can be used to work out the cost, in pounds, of hiring a floor sander.

Multiply the number of days by 25 and then add 50.

Betty hires a floor sander for d days.

The cost is £C.

Write down a formula for C in terms of d.

7 This rule can be used to change a distance given in miles, m, to a distance in kilometres, k.

First multiply by 8 then divide by 5.

  a Write this rule as a formula.

  b Use your formula to change 200 miles into kilometres.

8 A square has side length l cm.

Write down a formula for the area A of the square.

9 **a** Write down a formula for the perimeter P of the rectangle in terms of a.

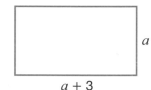

  b Work out the value of P when $a = 6$ cm.

10 $S = \frac{D}{T}$

Work out the value of D when $T = 6.5$ and $S = -10$.

11 Rearrange each formula to make the letter in square brackets the subject.

  a $v = u + at$ $[a]$

  b $V = \frac{1}{3} Ah$ $[h]$

  c $y = 3(x - 3)$ $[x]$

  d $v^2 = u^2 + 2as$ $[s]$

  e $s = \frac{1}{2} at^2$ $[t]$

  f $T = \sqrt{\frac{2s}{g}}$ $[g]$

  g $4ax = 3 + a$ $[a]$

Linear sequences

SNAPIT!   Sequences

1, 3, 5, 7, 9, 11, … is a
sequence of **odd numbers**.

2, 4, 6, 8, 10, 12, … is a
sequence of **even numbers**.

Number sequences

A pattern of numbers that follow a rule is called a **sequence**.

Each number in a sequence is called a **term**.

In the sequence 3, 6, 9, 12, … the first term is 3, the second term is 6, the third term is 9, and so on.

You can continue a sequence of numbers if you know the **term-to-term rule**. This rule tells you what to do to each term to obtain the next term in the sequence.

WORKIT!

Here are the first four terms in a sequence:

46, 42, 38, 34

a Write down the term-to-term rule of this sequence.

The term-to-term rule is subtract 4.

b Write down the next two terms in this sequence.

30, 26

c Write down the 10th term in this sequence.

7th term = 22

8th term = 18

9th term = 14

10th term = 10

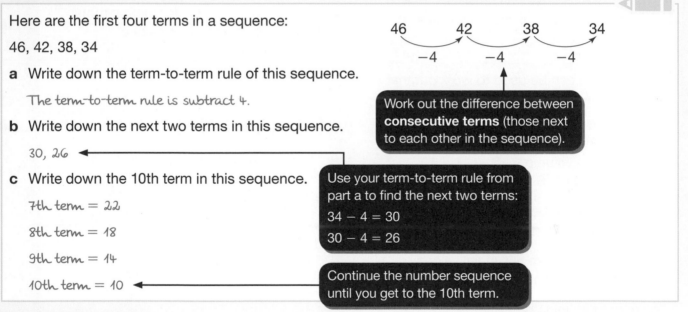

Work out the difference between **consecutive terms** (those next to each other in the sequence).

Use your term-to-term rule from part a to find the next two terms:

$34 - 4 = 30$

$30 - 4 = 26$

Continue the number sequence until you get to the 10th term.

You can also continue a number sequence, or find a specific term, if you know the **position-to-term rule**. This rule tells you what to do to the term number to obtain that term in the sequence.

WORKIT!

The position-to-term rule for a sequence is 'multiply the term number by 2 and add 4'.

a Find the first two terms of this sequence.

1st term = $1 \times 2 + 4 = 6$ ◄——— The 1st term is term number 1, the 2nd term is term number 2.

2nd term = $2 \times 2 + 4 = 8$

The first two terms are 6, 8.

b Find the 100th term of this sequence.

100th term = $100 \times 2 + 4 = 204$

Pattern sequences

Here are the first four patterns in a sequence of patterns made with squares.

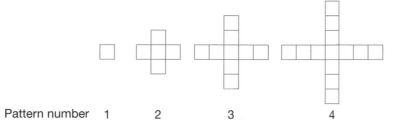

Pattern number   1   2   3   4

Pattern number	Number of squares
1	1
2	5
3	9
4	13

(+4 between each term)

If you want to find the number of squares in pattern number 8 you could draw patterns 5 to 8. Or you could use a table to record the number of squares used in each pattern (see left).

Using the table, the term-to-term rule is 'add 4'.

So the number sequence is 1, 5, 9, 13, 17, 21, 25, 29, ...

Since 29 is the eighth term in this sequence, there are 29 squares in pattern number 8.

The nth term

In an **arithmetic sequence** the terms increase or decrease by a fixed number called the **common difference**.

1, 4, 7, 10, ... is an arithmetic sequence with a common difference of 3.

27, 22, 17, 12, ... is an arithmetic sequence with a common difference of −5.

You need to be able to find the **nth term** of an arithmetic sequence. When the position-to-term rule of a sequence is given in algebraic form it is called the nth term.

NAIL IT!

The terms in a sequence are numbered 1, 2, ... n.

The nth term is the term at position n in the sequence.

WORKIT!

Here are the first four terms of an arithmetic sequence.

5, 9, 13, 17

Find an expression, in terms of n, for the nth term of this sequence.

Term number:	1	2	3	4	...	n
Term:	5	9	13	17		
Common difference:		+4	+4	+4		+1
4 × term number:	4	8	12	16	...	$4n$

The nth term is $4n + 1$

To work out the nth term:

1. Find the common difference: + 4

2. Multiply each term number by the common difference:
$1 \times 4 = 4, 2 \times 4 = 8, 3 \times 4 = 12, 4 \times 4 = 16, ... 4n$

3. Work out how to get from each term in $4n$ to the term in the original sequence: + 1

You should always check your expression for the nth term by substituting values of n into your expression.

1st term when $n = 1$: $4n + 1 = 4 \times 1 + 1 = 5$ ✓

2nd term when $n = 2$: $4n + 1 = 4 \times 2 + 1 = 9$ ✓

You can also use the nth term to find any term of the sequence.

For the 50th term: $n = 50$, $4n + 1 = 4 \times 50 + 1 = 201$

You may need to decide whether a given number is a term in the sequence, or find the first term that is greater or less than a certain number.

WORKIT!

The nth term of the arithmetic sequence 1, 3, 5, 7, … is $2n - 1$.

a Is 86 a number in this sequence?

$86 = 2n - 1$ ← Write an equation using the nth term. Solve it for n.

Add 1 to both sides. → $87 = 2n$

$43.5 = n$ ← Divide both sides by 2.

The value of n has to be a positive whole number if the number is a term in a sequence. → 86 cannot be in the sequence because 43.5 is not an integer.

b Find the first term in the sequence that is greater than 100.

$100 = 2n - 1$

$101 = 2n$

$50.5 = n$

The 51st term is the first term that is greater than 100:

$2n - 1 = 2 \times 51 - 1 = 101$

DOIT!

1 The nth term of a sequence is $2n + 3$.
Which term has a value of 79?

2 The arithmetic sequence 13, 10, 7, 4, 1, … gets smaller and smaller.
Show that the nth term of this sequence is $16 - 3n$.

✓ CHECKIT!

1 For each sequence

  a Write down the next two terms.

  b Write down the 10th term.

  i 2, 5, 8, 11, …   **iii** 3, 9, 15, 21, …

  ii 23, 19, 15, 11, …   **iv** 4, 9, 14, 19, …

2 The position-to-term rule for a sequence is 'multiply by 4 and subtract 2'.

  a Find the first four terms of this sequence.

  b Find the 20th term of this sequence.

3 a Here are the first three terms of a sequence.

  $-25, -18, -11$

  Find the first two terms in the sequence that are greater than zero.

  b The nth term of a different sequence is $15 - 2n$.

  Which term is the first one with a negative value?

4 Here are the first three patterns in a sequence. The patterns are made from squares and triangles.

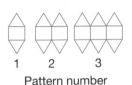

1   2   3
Pattern number

  a Draw pattern number 4.

  b Find the number of triangles in pattern number 8.

  c Kayleigh says 'One of the patterns will have 35 triangles'.

  Is Kayleigh correct? Give a reason for your answer.

5 Here are the first five terms of an arithmetic sequence:

  3, 7, 11, 15, 19

  a Find an expression, in terms of n, for the nth term of this sequence.

  b Is 99 a term in the sequence? Give a reason for your answer.

Non-linear sequences

SNAPIT! Special sequences to remember

Sequence	
Triangular numbers 1, 3, 6, 10, 15, 21, ...	Generated from a pattern of dots that form a triangle:
Square numbers 1, 4, 9, 16, 25, 36, 49, ...	The sequence of square numbers: 1^2, 2^2, 3^2, 4^2, 5^2, 6^2, ... This is also an example of a quadratic sequence.
Cube numbers 1, 8, 27, 64, 125, 216, ...	The sequence of cube numbers: 1^3, 2^3, 3^3, 4^4, 5^3, 6^3, ...
Fibonacci-type 1, 1, 2, 3, 5, 8, ...	The next term in the sequence is found by adding the previous two terms together.
Quadratic 2, 5, 10, 17, 26, 37, ...	Has n^2 and no higher power of n in its nth term. For the example, the nth term is $n^2 + 1$
Geometric sequence (or geometric progression) 1, 3, 9, 27, 81, 243, ...	The next term is found by multiplying the previous term by a fixed number. For the example, the term-to-term rule is 'multiply by 3'.

To find the term-to-term rule of a geometric sequence you need to work out what each term is multiplied by.

WORKIT!

Here are the first three terms in a geometric sequence.

0.001, 0.01, 0.1

a Write down the term-to-term rule of this geometric sequence.

0.001   0.01   0.1

× 10   × 10

The term-to-term rule is multiply by 10.

b Write down the next two terms in this geometric sequence.

1, 10

SNAPIT!

Square numbers

You need to know these square numbers:

$1^2 = 1$	$6^2 = 36$	$11^2 = 121$
$2^2 = 4$	$7^2 = 49$	$12^2 = 144$
$3^2 = 9$	$8^2 = 64$	$13^2 = 169$
$4^2 = 16$	$9^2 = 81$	$14^2 = 196$
$5^2 = 25$	$10^2 = 100$	$15^2 = 225$

Use your term-to-term rule from part **a** to work out the next two terms:

$0.1 \times 10 = 1$

$1 \times 10 = 10$

To generate terms in a quadratic sequence you substitute term numbers into a given nth term which involves an n^2 term.

One 'term' in a sequence can be a result of a combination of algebraic terms. For example, $3n^2$ and $2n$ are both algebraic terms. They could combine to form one 'term' in a sequence: $3n^2 + 2n$.

NAILIT!

A geometric sequence could have negative, fraction or decimal terms.

WORKIT!

The nth term of a sequence is $2n^2$.
Work out the first two terms and the 10th term of this sequence.

1st term when $n = 1$: $2n^2 = 2 \times 1^2 = 2$
2nd term when $n = 2$: $2n^2 = 2 \times 2^2 = 8$ ← Only the value of n is squared: $2n^2$ is not the same as $(2n)^2$.
10th term when $n = 10$: $2n^2 = 2 \times 10^2 = 200$

DOIT!

Triangular numbers

Write down the pattern of common differences for the triangular number sequence
1, 3, 6, 10, 15, 21, …

What can you say about the common difference?

CHECKIT!

1 Describe each of these number sequences as either a geometric, Fibonacci-type, square number or arithmetic sequence.

1, 3, 5, 7, 9, …   1, 4, 5, 9, 14, …
1, 2, 4, 8, 16, …   1, 4, 9, 16, 25, …

2 Write down the next two terms in the following quadratic sequence.
3, 6, 11, 18, …

3 Write down the next two terms in each geometric sequence.

a 4, 2, 1, …   d $\frac{1}{9}, \frac{1}{3}, 1, …$

b 5, 0.5, 0.05, …   e −0.1, −0.2, −0.4, …

c $\frac{1}{2}, \frac{1}{4}, \frac{1}{8}, …$ f 3, −6, 12, …

4 A Fibonacci-type sequence is made by adding the previous two terms together.

The first five terms are 2, 2, 4, 6, 10.
Work out the 8th term.

5 Write down the first four terms of a sequence where the nth term is given by $n^2 + 5$.

6 The nth term of a sequence is $3n^2 - 4$.
Work out the 5th term of this sequence.

7 The nth term of a sequence is $n^2 + 2n$.
Work out the first three terms of this sequence.

8 Here are the first three terms of a Fibonacci-type sequence.

$a, b, a + b$

a Show that the 5th term of this sequence is $2a + 3b$.

The 2nd term of the sequence is 5.
The 5th term of the sequence is 23.

b Work out the value of a.

Show that...

$\frac{10x}{2} \equiv 5x$ is an **identity**. An identity is true for all values of the letters. An identity can be written with the identity symbol '$\equiv$' or with an equals sign.

$2(n + 1) = 2n + 2$ is also an identity. You can show this by expanding the brackets on the left-hand side (LHS) to show that this expression is identical to the one on the right-hand side (RHS).

LHS $= 2(n + 1) \equiv 2n + 2$ RHS $= 2n + 2$

To show that expressions are equivalent, you need to use algebra − in particular, expanding brackets (page 43) and simplifying expressions (page 39).

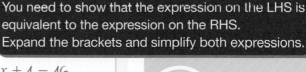

SNAPIT!

Symbols

'$=$' means 'is equal to'
  $x + 2 = 6$ when $x = 4$
'$\neq$' means 'is **not** equal to'
  $2 + 2 \neq 5$
'$\equiv$' means 'is identically equal to'
  $2(n + 1) \equiv 2n + 2$

WORKIT!

Show that $(x - 1)^2 - 16 \equiv (x + 3)(x - 5)$ ◄

> You need to show that the expression on the LHS is equivalent to the expression on the RHS.
> Expand the brackets and simplify both expressions.

LHS $= (x - 1)^2 - 16 \equiv (x - 1)(x - 1) - 16 = x^2 - x - x + 1 - 16$
$= x^2 - 2x - 15$

RHS $= (x + 3)(x - 5) = x^2 - 5x + 3x - 15$
$= x^2 - 2x - 15$

LHS $=$ RHS
So $(x - 1)^2 - 16 \equiv (x + 3)(x - 5)$

NAILIT!

When you are told to 'Show that …', write down **every** step of your working.

You also need to be able to show a geometric result.

WORKIT!

The diagram shows a quadrilateral and a triangle.
Show that the perimeter of the quadrilateral is 2 times the perimeter of the triangle.

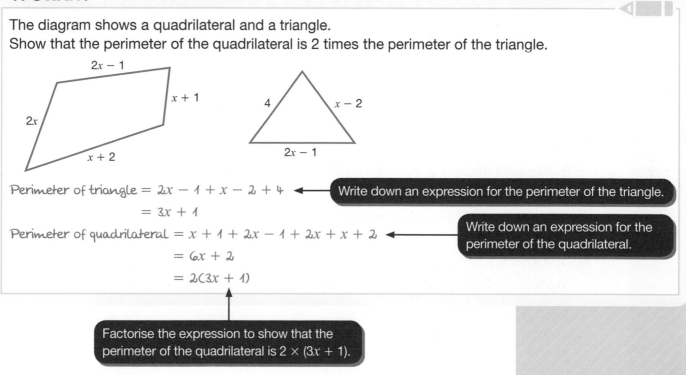

Perimeter of triangle $= 2x - 1 + x - 2 + 4$ ◄ Write down an expression for the perimeter of the triangle.
$= 3x + 1$

Perimeter of quadrilateral $= x + 1 + 2x - 1 + 2x + x + 2$ ◄ Write down an expression for the perimeter of the quadrilateral.
$= 6x + 2$
$= 2(3x + 1)$

Factorise the expression to show that the perimeter of the quadrilateral is $2 \times (3x + 1)$.

You also need to have a knowledge of odd and even numbers.

If n is an integer, then $2n$ is even and $2n + 1$ is odd.

For example, if $n = 5$, $2n = 2 \times 5 = 10$ (even) and $2n + 1 = 2 \times 5 + 1 = 11$ (odd).

Note that $2n - 1$ is also odd: $2 \times 5 - 1 = 9$.

When you multiply an odd number by an even number, the answer is always even. You could use any of the following examples to test this: $3 \times 2 = 6$, $5 \times 10 = 50$, $11 \times 4 = 44$.

NAILIT!

An integer is a positive or negative whole number.

STRETCHIT!

Odd or even?

If n is an **even** number,

explain why $(n - 1)(n + 1)$ will always be odd.

Let m be an **odd** number.

Explain why $(m - 1)(m + 1)$ will always be even.

CHECKIT!

1 a Luca says that the sum of two prime numbers is always odd.

Write down an example to show that Luca is incorrect.

b Mo says that if n is an even number then $\frac{n}{2}$ is always an even number.

Is Mo correct? Give a reason for your answer.

2 Show that
a $(x + 2)(x - 2) \equiv x^2 - 4$
b $(x - 3)^2 \equiv x^2 - 6x + 9$
c $(x + 1)^2 + 4 \equiv x^2 + 2x + 5$
d $6(a - 3) - 2(2a - 5) + 6 \equiv 2(a - 1)$
e $4(x - 3) + 2(x + 5) \equiv 3(2x - 1) + 1$

3 Show that the values of a and b in the identity are 2 and 1 respectively

$4(ax - 2) + 5(3x + b) \equiv 23x - 3$

4 Here are three rods.
Rod A has length n cm. Rod B is 1 cm longer than rod A. Rod C is 2 cm longer than rod A.
Show that the total length of rods A and C is 2 times the length of rod B.

Functions

A **function** is a rule for calculating one value by doing something to another value. These values are often called y and x.

A **function machine** (or **number machine**) is one way of showing this. It has an input, an output and one or more operations. The operation $(+, -, \times, \div)$ tells you what to do to the input to obtain the output.

This is a two-stage function machine, as there are two operations to be performed.

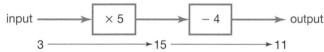

If you know the output, you can work out the input by working backwards.

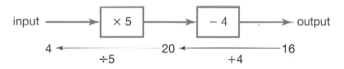

You can use a function machine to produce a **sequence** of numbers.

WORKIT!

Here is a table for a two-stage function machine. It multiplies by 2 and then subtracts 3.

a Complete the table.

b The term number is n.

Write down an expression, in terms of n, for the nth term.

$2n - 3$

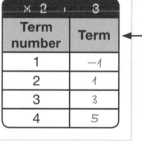

Term number	Term
1	−1
2	1
3	3
4	5

> The 'term number' is the 'input' and the 'term' is the 'output'.
>
> For term number 1 the 'input' is 1.
>
> Working:
> $1 \times 2 - 3 = -1$ ('output')
>
> For term number n,
> $n \times 2 - 3 = 2n - 3$
>
> See page 64 for more on sequences.

$y = 2x$ and $y = 3x + 2$ are both examples of **functions**. A function is the rule that tells you how to work out values of y for given values of x.

By using the values of x as the input, a function machine can be used to find the output values y. This will give you a set of coordinate pairs (x, y).

> This is a useful skill for when you need to plot graphs (see page 75).

The function machine below represents the function $y = 2x - 4$ ('multiply x by 2, then subtract 4').

WORKIT!

Here is a function machine.
Work out the output y when:

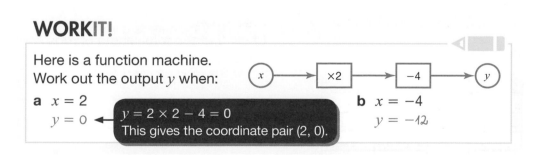

a $x = 2$

$y = 0$

$y = 2 \times 2 - 4 = 0$

This gives the coordinate pair (2, 0).

b $x = -4$

$y = -12$

For the function machine below, in terms of the input x, the output y is:
$y = 5x - 4$ ('multiply x by 5, then subtract 4').

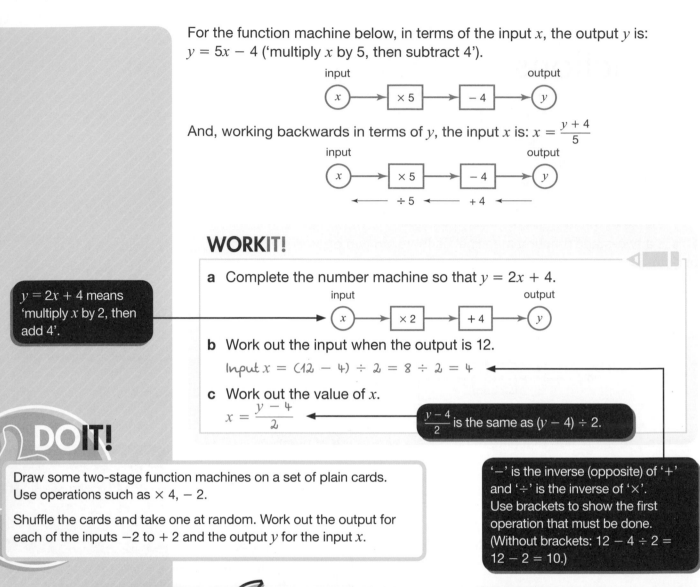

And, working backwards in terms of y, the input x is: $x = \dfrac{y + 4}{5}$

WORKIT!

a Complete the number machine so that $y = 2x + 4$.

> $y = 2x + 4$ means 'multiply x by 2, then add 4'.

b Work out the input when the output is 12.

Input $x = (12 - 4) \div 2 = 8 \div 2 = 4$

c Work out the value of x.

$x = \dfrac{y - 4}{2}$

> $\dfrac{y - 4}{2}$ is the same as $(y - 4) \div 2$.

DOIT!

Draw some two-stage function machines on a set of plain cards. Use operations such as $\times 4, - 2$.

Shuffle the cards and take one at random. Work out the output for each of the inputs -2 to $+2$ and the output y for the input x.

> '$-$' is the inverse (opposite) of '$+$' and '$\div$' is the inverse of '$\times$'.
> Use brackets to show the first operation that must be done.
> (Without brackets: $12 - 4 \div 2 = 12 - 2 = 10$.)

CHECKIT!

1 Here is a function machine.

Work out the output y when

a $x = 2$

b $x = 1$

c $x = -4$

2 Complete the table for each function.

a

$\times 1 \rightarrow + 2$	
x	y
-2	
0	
2	4

b

$\times 2 \rightarrow - 3$	
x	y
-2	
0	-3
2	

3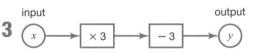

a Write down the output y in terms of x.

b Work out the output when the input is 10.

c Work out the value of x in terms of y.

d Sasha says 'There is a value of x where it is equal to y'.

Show that Sasha is correct.

Coordinates and midpoints

Coordinates

Coordinates are used to describe the position of a point on a coordinate grid. The first number in the coordinate pair (x, y) gives the horizontal position; the second number (x, y) gives the vertical position.

Coordinates can include positive and negative numbers.

WORKIT!

Here is a grid showing points P, Q and R.

a   Write down the coordinates of the points P and R.

    P (−1, 4)   R (1, 1)

b   On the grid, mark with a cross the point S so that PQRS is a parallelogram.

Opposite sides of a parallelogram are equal length and parallel. Plot point S so that it is level with R and one square to the left of point P.

DOIT!

Use your knowledge of 2D shapes:
Draw a coordinate grid with axes from −8 to +8.
On your grid, draw a square, rhombus, rectangle, parallelogram, kite and triangle.

STRETCHIT!

A is the point (2, 4). B is the point (−1, 3).

B is the midpoint of AC.

Work out the coordinates of C.

Midpoint of a line

The **midpoint** of a line is exactly halfway along the line.

To find the midpoint, add the x coordinates and divide by 2 and add the y coordinates and divide by 2.

Add the x coordinates and divide by 2.

WORKIT!

Find the coordinates of the midpoint of AB.

A(−3, 0), B(4, 2)

x coordinate:

4 + (−3) = 1   1 ÷ 2 = 0.5

y coordinate:

2 + 0 = 2   2 ÷ 2 = 1

Midpoint is (0.5, 1)

Add the y coordinates and divide by 2.

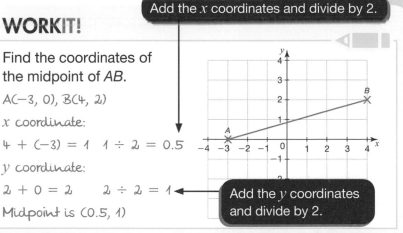

✓ CHECKIT!

1 Here is a grid showing the points A and B.

   a   Write down the coordinates of the point A.

   b   On the grid, mark with a cross the point (−4, −3). Label this point C.

   c   On the grid, mark with a cross a point D so that the quadrilateral ABCD is a rectangle.

   d   Find the coordinates of the midpoint of BD.

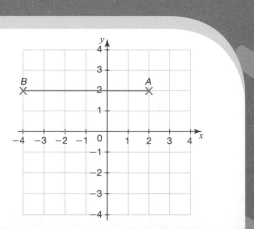

Straight-line graphs

Drawing straight-line graphs of the form $y = mx + c$

SNAP IT! Horizontal and vertical lines

A horizontal line on a graph has equation $y = a$, where a is a number.

A vertical line on a graph has equation $x = a$, where a is a number.

The x-axis has equation $y = 0$.

The y-axis has equation $x = 0$.

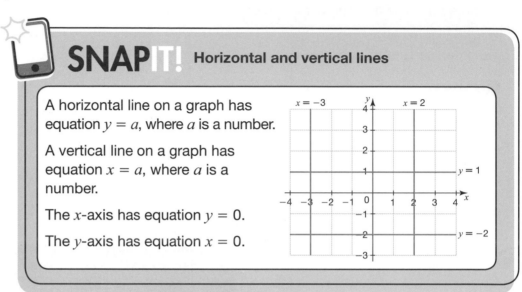

The equation of a straight line on a graph is $y = mx + c$ where m and c are numbers.

m is the **gradient**. This tells you how steep the line is.

Lines that slope upwards from left to right have positive gradients.

Lines that slope downwards from left to right have negative gradients.

c is the **y-intercept**. This is the point where the line crosses the y-axis. The coordinates of this point are $(0, c)$.

SNAP IT! Graphs of $y = x$ and $y = -x$

The straight line $y = x$ has gradient 1 and y-intercept $(0, 0)$.

The straight line $y = -x$ has gradient -1 and y-intercept $(0, 0)$.

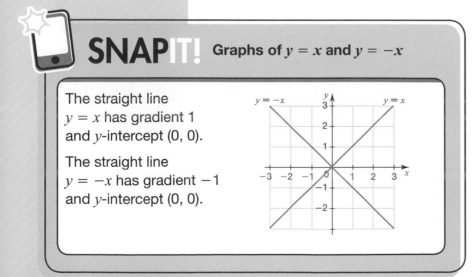

You can compare the equation of a straight line with the general form of the equation $y = mx + c$ to identify the gradient and y-intercept.

The straight line $y = 2x - 4$ has gradient 2 ($m = 2$) and y-intercept $(0, -4)$ ($c = -4$).

The straight line $y = -3x + 1$ has gradient -3 ($m = -3$) and y-intercept $(0, 1)$ ($c = 1$).

To plot a straight-line graph you need to:

1 Make a table of some values of x. Substitute these into the equation to find the values of y.

2 Plot the points from the table on the coordinate grid.

3 Join the points with a straight line.

WORKIT!

Draw the graph of $y = 2x - 1$ for values of x from -1 to 3.

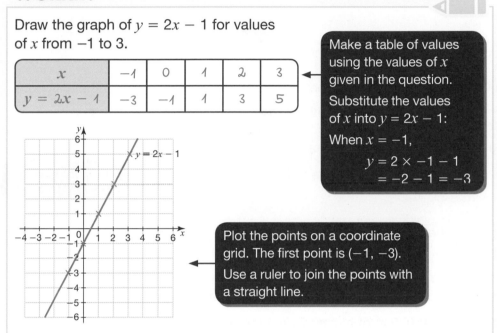

x	-1	0	1	2	3
$y = 2x - 1$	-3	-1	1	3	5

Make a table of values using the values of x given in the question.

Substitute the values of x into $y = 2x - 1$:

When $x = -1$,
$$y = 2 \times -1 - 1$$
$$= -2 - 1 = -3$$

Plot the points on a coordinate grid. The first point is $(-1, -3)$.

Use a ruler to join the points with a straight line.

You can sketch a straight-line graph using the gradient and y-intercept.

WORKIT!

On the grid, draw the graph of $y = -2x + 1$

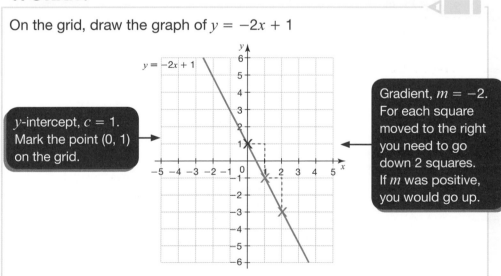

y-intercept, $c = 1$.
Mark the point $(0, 1)$ on the grid.

Gradient, $m = -2$. For each square moved to the right you need to go down 2 squares. If m was positive, you would go up.

Finding the equation of a straight line

The **gradient** of a line is a measure of its slope. You can find the gradient of a straight line by drawing a right-angled triangle on the line.

$$\text{Gradient} = \frac{\text{difference in } y \text{ coordinates}}{\text{difference in } x \text{ coordinates}}$$

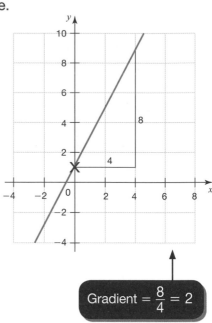

NAILIT!

To work out the gradient of a straight line:

- Draw a large triangle with a base that is a whole number of units.
- Use the scales on the axes to work out the differences in y and x coordinates.

Gradient $= \dfrac{8}{4} = 2$

To write the equation of a line in the form $y = mx + c$ you need to know the gradient (m) and the y-intercept (c).

For the example above, you found that the gradient $m = 2$. The y-intercept can be read off the graph: $c = 1$. So the equation of this line is $y = 2x + 1$.

To find the equation of a straight line through two given points, always draw a sketch to show the two points.

WORKIT!

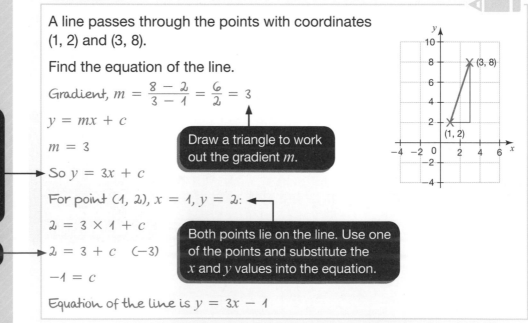

A line passes through the points with coordinates (1, 2) and (3, 8).

Find the equation of the line.

Gradient, $m = \dfrac{8 - 2}{3 - 1} = \dfrac{6}{2} = 3$

$y = mx + c$

$m = 3$

Draw a triangle to work out the gradient m.

So $y = 3x + c$

Write down the general form of the equation of a straight line: $y = mx + c$.

Substitute the value of m into the equation.

For point (1, 2), $x = 1$, $y = 2$:

$2 = 3 \times 1 + c$

Both points lie on the line. Use one of the points and substitute the x and y values into the equation.

Solve the equation for c.

$2 = 3 + c$ (−3)

$-1 = c$

Equation of the line is $y = 3x - 1$

You also need to be able to find the equation of a straight line through one point with a given gradient. Since you are given the gradient m you simply need to work through the last part of the example above to find c.

More straight-line graphs

Lines of the form $ax + by = c$

When the equation of a straight line is given in the form $ax + by = c$, find the x- and y-intercepts to draw its graph.

> Use $x = 0$ to find one point:
> $2 \times 0 + y = 4$
> $y = 4$
> Use $y = 0$ to find the second point:
> $2x + 0 = 4$
> $2x = 4$ $(\div 2)$
> $x = 2$

WORKIT!

Draw the graph of $2x + y = 4$.

x	0	2
y	4	0

> Plot the two points:
> (0, 4), (2, 0).
> Join the points
> with a straight line.

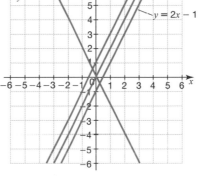

Gradients and parallel lines

The gradient of a straight line is a measure of how steep it is. Steeper lines have larger gradients.

The line $y = 4x - 2$ has a steeper gradient ($m = 4$) than the line $y = 2x + 4$ ($m = 2$).

Parallel lines have the same gradient. Three of these lines are parallel.

They have a gradient of 2.

$y = 2x$ $y = 2x + 1$ $y = 2x - 1$

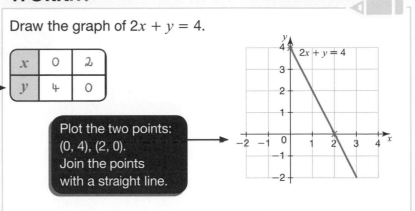

The line $y = -2x$ is not parallel. This line has a gradient of -2.

Parallel lines can be identified by looking at their equations. You may first need to rearrange the equation into the form $y = mx + c$.

> If you're unsure about changing the subject of an equation, have a look at page 61.

WORKIT!

Here are the equations of four straight lines.

A $y = 3x + 1$ B $1 - 3x = y$ C $y - 3x = 4$ D $3y = x + 1$

Two of the lines are parallel.

Write down the two parallel lines.

A: $y = 3x + 1$ ← The equation is already in the form $y = mx + c$

B: $1 - 3x = y$

    $y = -3x + 1$ ← Rewrite the equation in the form $y = mx + c$

C: $y - 3x = 4$ $(+ 3x)$

    $y = 4 + 3x$

    $y = 3x + 4$ ← Rearrange the equation into the form $y = mx + c$

D: $3y = x + 1$ $(\div 3)$

    $y = \frac{1}{3}x + \frac{1}{3}$ ← Rearrange the equation into the form $y = mx + c$

Lines A and C are parallel.   ← Parallel lines have the same gradient. Lines A and C have gradient 3. Make sure you don't ignore the negative signs. Line B has gradient -3. This means the line slopes downwards from left to right.

y-intercepts

You may also need to rearrange the equation of a straight line into the form $y = mx + c$ to identify the *y*-intercept (c).

For example:

$$3y = x + 6$$

> Divide both sides by 3.

$$y = \frac{x}{3} + 2$$

The *y*-intercept is (0, 2).

Using graphs to solve linear equations

> Solving linear equations algebraically is covered on page 51.

You can use graphs to solve linear equations such as $2x + 1 = 0$ and $3x - 2 = 3$. This is a **graphical method**.

You solve linear equations such as $2x + 1 = 0$ by reading off the *x* coordinate where the graph $y = 2x + 1$ crosses the *x*-axis ($y = 0$).

WORKIT!

> Substitute the values of x into $y = 2x + 1$
>
> When $x = -2$,
> $y = 2 \times (-2) + 1$
> $\quad = -4 + 1 = -3$

a Complete the table of values for $y = 2x + 1$

x	−2	−1	0	1	2
y	−3	−1	1	3	5

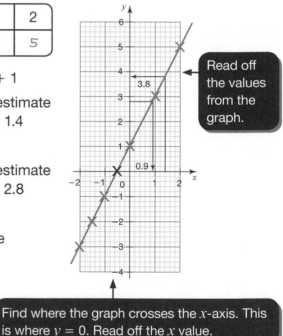

> Read off the values from the graph.

b Draw the graph of $y = 2x + 1$

c Use your graph to find an estimate for the value of y when $x = 1.4$

When $x = 1.4$, $y = 3.8$

d Use your graph to find an estimate for the value of x when $y = 2.8$

When $y = 2.8$, $x = 0.9$

e Use your graph to solve the equation $2x + 1 = 0$

$x = -0.5$

> Find where the graph crosses the *x*-axis. This is where $y = 0$. Read off the *x* value.

You solve linear equations such as $3x - 2 = 3$ by reading off the *x* coordinate at the point of intersection of the graph $y = 3x - 2$ with the graph $y = 3$.

STRETCHIT!

> Use the graph of $y = 3x - 2$ on the next page to solve the equation $3x - 2 = -x$.

WORKIT!

Here is the graph of
$y = 3x - 2$.

Use the graph to solve
the equation $3x - 2 = 3$.

$x \approx 1.7$

Read off the x value as
accurately as possible.
The $\approx$ sign means
'approximately equal to'.

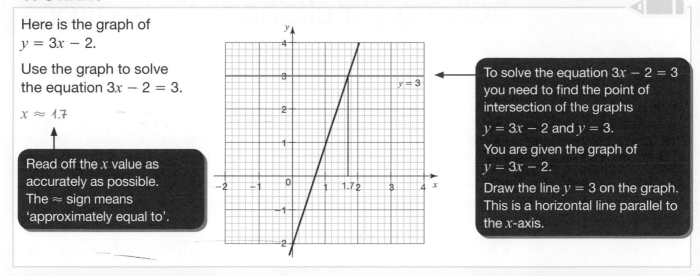

To solve the equation $3x - 2 = 3$
you need to find the point of
intersection of the graphs
$y = 3x - 2$ and $y = 3$.
You are given the graph of
$y = 3x - 2$.
Draw the line $y = 3$ on the graph.
This is a horizontal line parallel to
the x-axis.

CHECKIT!

1 Draw the graph of $y = -x + 2$ for
values of x from -2 to 2.

2 Draw the graph of $y = 2x - 3$.

3

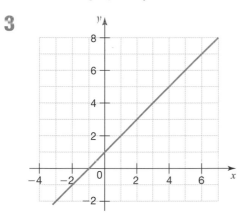

Find the equation of the straight line.

4 A straight line has gradient 2.
The point $(1, -2)$ lies on the line.

Find the equation of the line.

5 A straight line passes through the points
with coordinates $(0, 4)$ and $(4, 2)$.

Find the equation of the line.

One of the points is $(0, 4)$ so you
can write down the y-intercept
of the line without calculation.

6 Draw the graphs of

  a $x + y = 6$ **b** $2y + x = 4$

7 Here are the equations of four
straight lines

  A $y = 4x + 1$ C $x - 2y = 2$

  B $4x + 4y = 4$ D $2y = 4 + 8x$

Write down the two lines that are parallel.

8 Write down the y-intercept of the straight
line with equation $2y = x - 4$.

9 Here is the graph of
$y = 2x - 1$

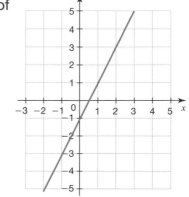

  a Use the graph to find an estimate
for y when $x = 2.2$.

  b Use the graph to solve the equation
$2x - 1 = 0$

  c Use the graph to solve the equation
$2x - 1 = 4$

  d Use the graph to solve the equation
$2x - 1 = -2$

79

Solving simultaneous equations

Algebraic solution

When there are two unknowns you need two equations. These are called **simultaneous equations**. You need to find the values for the two unknowns which make **both** equations true.

Simultaneous equations can be solved using the substitution or elimination method. For both methods you need to be able to solve linear equations.

Substitution method

WORKIT!

Solve the simultaneous equations $y = x + 4$ (1)

Number each equation. ➝ $3x + y = 8$ (2)

Substitute equation (1) into equation (2): $3x + x + 4 = 8$

Collect like terms. ➝ $4x + 4 = 8$

Subtract 8 from both sides. ➝ $4x = 4$

Divide both sides by 4. ➝ $x = 1$

Substitute $x = 1$ into equation (1): $y = 1 + 4$

$y = 5$

Solution is $x = 1$, $y = 5$

Check: $3x + y = 3 \times 1 + 5 = 3 + 5 = 8$ ✔

DOIT!

Solve the simultaneous equations

$x + 3y = 9$
$y = 2x - 4$

NAILIT!

Check your answer by substituting both values into the original equation that you haven't used.

Elimination method

WORKIT!

Solve the simultaneous equations

$2x + y = 5$ (1)
$x - 3y = 6$ (2)

$3 \times (1)$: $6x + 3y = 15$ (3) ◄

$(3) + (2)$: $7x = 21$ $(\div 7)$

$x = 3$

Substitute $x = 3$ into (1):

$2 \times 3 + y = 5$

$6 + y = 5$ $(- 6)$

$y = -1$

Solution is $x = 3$, $y = -1$

Check: $x - 3y = 3 - 3(-1) = 3 + 3 = 6$ ✔

The signs of the terms in y are different so choose this unknown to eliminate.

Multiply equation (1) by 3 to make the coefficients of y (the numbers in front of the ys) equal.

NAILIT!

When deciding which unknown to eliminate, choose the unknown where the signs are different if possible. Then you can add the equations to eliminate the unknown.

For these simultaneous equations, both equations must be multiplied by a number before an unknown can be eliminated.

$2x + 4y = 10$ (1)

$3x + 3y = 12$ (2)

The signs of the terms in both x and y are the same in both equations. Choose to eliminate x.

$6x + 12y = 30$ (3)

$6x + 6y = 24$ (4)

Multiply (1) by 3 and (2) by 2 to make the coefficients of x equal.

$6y = 6$

Subtract (4) from (3) to eliminate the terms in x.

$y = 1$

Divide both sides by 6.

$2x + 4 = 10$

Substitute $y = 1$ into (1) to find the value of x.

$2x = 6$

Subtract 4 from both sides.

$x = 3$

Divide both sides by 2.

Solution is $x = 3$, $y = 1$

Setting up simultaneous equations

You can set up simultaneous equations to represent real-life situations and solve problems.

You need to define the unknowns, then turn each sentence in the problem into an equation.

WORKIT!

Two coffees and a muffin cost £3.75.

Three coffees and two muffins cost £6.00.

Find the cost of a coffee and the cost of a muffin.

Define the unknowns. You can use any letters to represent them.

Let c = coffee and m = muffin.

$2c + m = 375$ (1)

$3c + 2m = 600$ (2)

Turn each sentence into an equation. It is easier to work in pence.

$2 \times (1): 4c + 2m = 750$ (3)

$(3) - (2): c = 150$

Once the equations have been set up, solve the simultaneous equations as usual.

Substitute $c = 150$ into (1): $2 \times 150 + m = 375$

$300 + m = 375$

$m = 75$

Subtract 300 from both sides.

Check: $3c + 2m = 3 \times 150 + 2 \times 75 = 450 + 150 = 600$ ✔

A coffee costs £1.50.

A muffin costs 75p.

Make sure you answer the question asked by stating the cost of a coffee and the cost of a muffin.

DO IT!

Sometimes the coefficients of the terms in x or the terms in y are equal so you do not need to multiply before you can eliminate an unknown.

Solve the simultaneous equations

$2x + y = 7$ (1)

$x - y = 5$ (2)

by adding (1) and (2) to eliminate the terms in y.

NAIL IT!

When setting up simultaneous equations to represent a real-life situation, make sure that both sides of each equation use the same units.

Graphical solution

You can solve simultaneous equations by drawing the graphs of the two equations and finding the point of intersection. If you're unsure about drawing straight-line graphs, have a look at page 74.

WORKIT!

Solve the simultaneous equations

$x + 2y = 4$

$x - y = 1$

For each equation, plot the two points and join with a straight line.

Find the x-intercept $(y = 0)$ and y-intercept $(x = 0)$ of both equations.

$x + 2y = 4$

x	0	4
y	2	0

$x - y = 1$

x	0	1
y	−1	0

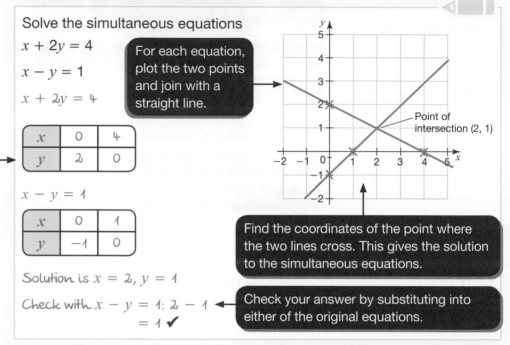

Point of intersection (2, 1)

Find the coordinates of the point where the two lines cross. This gives the solution to the simultaneous equations.

Solution is $x = 2, y = 1$

Check with $x - y = 1$: $2 - 1$

$\qquad\qquad\qquad = 1$ ✔

Check your answer by substituting into either of the original equations.

CHECKIT!

1 $x + y = 16$

$x - y = 5$

Work out the values of x and y.

2 Solve these simultaneous equations.

a $2x + y = 4$
$\quad 3x - y = 1$

b $x - y = 5$
$\quad 2x + y = 4$

c $2x + y = 8$
$\quad x + y = 2$

d $4x - y = 10$
$\quad x + 2y = 7$

e $2x + y = 7$
$\quad x - 4y = 8$

f $2x + 3y = 7$
$\quad 3x - 2y = 4$

3 The sum of two numbers is 21 and their difference is 7.

Find the value of each number.

Let the two numbers be x and y.

4 3 burgers and 2 colas cost £5.05.

3 burgers and 4 colas cost £7.25.

Find the cost of a burger and the cost of a cola.

5 Kirsty and Nathan are each booking tickets for a pantomime.

Kirsty buys 2 adult tickets and 4 child tickets. She pays £60.

Nathan buys 3 adult tickets and 3 child tickets. He pays £82.50.

Work out the cost of an adult ticket and the cost of a child ticket.

6 Use a graphical method to solve these simultaneous equations.

a $x + y = 10$
$\quad y - 3x = 2$

b $x - y = 7$
$\quad x + 4y = 12$

Quadratic graphs

A **quadratic function** contains an x^2 term but no higher power of x.

$y = x^2$, $y = 2x^2 + 1$, $y = 5 - x^2$ and $y = 4x^2 - 2x + 3$ are all examples of quadratic functions.

The graph of a quadratic function is a curved shape. A quadratic function with a positive x^2 term has a ∪-shaped curve. A quadratic function with a negative x^2 term has an upside-down ∪-shaped curve.

To understand what a function is, see page 71.

SNAP IT! Graphs of $y = x^2$ and $y = -x^2$

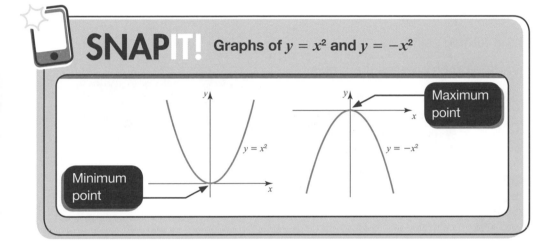

Minimum point

Maximum point

NAIL IT!

Drawing quadratic graphs

- Plot each point with a cross (×).
- Draw a smooth curve through the points.

The curve should pass through **all** the points!

The minimum or maximum point is called a **turning point** (or **stationary point**) as this is where the graph turns.

All quadratic graphs have a **line of symmetry**.

To draw a quadratic graph you need to complete a table of values.

DO IT!

Draw the graph of $y = 2x^2 + 2$ for values of x between -3 and $+3$.

WORK IT!

a Complete the table of values for $y = x^2 + 2$.

x	−3	−2	−1	0	1	2	3
y	11	6	3	2	3	6	11

b Draw the graph of $y = x^2 + 2$.

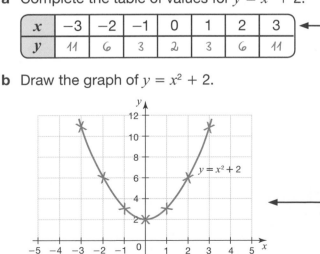

Substitute each value of x into $y = x^2 + 2$ to work out the corresponding value of y.

Put negative values of x in brackets and use the correct order of operations (**BIDMAS**).

When $x = -3$, $y = (-3)^2 + 2$
$= 9 + 2 = 11$

When $x = 2$, $y = 2^2 + 2$
$= 4 + 2 = 6$

Plot the points from your table on the graph: $(-3, 11)$, $(-2, 6)$…

Join your points with a **smooth** curve.

WORKIT!

a Draw the graph of $y = x^2 + 2x - 3$ for values of x from -4 to 2.

x	-4	-3	-2	-1	0	1	2
y	5	0	-3	-4	-3	0	5

Take care with the negative values.

When $x = -4$, $y = (-4)^2 + 2(-4) - 3$
$= 16 - 8 - 3$
$= 5$

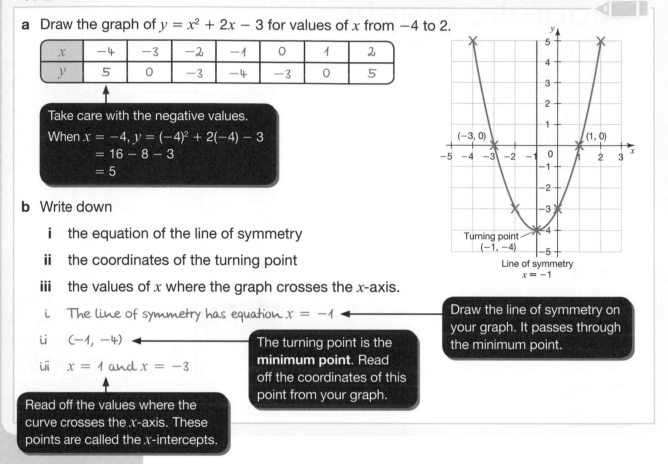

b Write down

i the equation of the line of symmetry

ii the coordinates of the turning point

iii the values of x where the graph crosses the x-axis.

i   The line of symmetry has equation $x = -1$

Draw the line of symmetry on your graph. It passes through the minimum point.

ii $(-1, -4)$

The turning point is the **minimum point**. Read off the coordinates of this point from your graph.

iii $x = 1$ and $x = -3$

Read off the values where the curve crosses the x-axis. These points are called the x-intercepts.

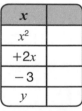

DOIT!

Copy and complete this table of values for $y = x^2 + 2x - 3$ for values of x from -4 to 2.

x	
x^2	
$+2x$	
-3	
y	

Compare this table with the one above in Work it! Which do you prefer?

Using graphs to solve quadratic equations

A **quadratic equation** contains an x^2 term but no higher power of x.

$x^2 - x - 4 = 0$ and $x^2 - 3x - 2 = -3$ are both examples of quadratic equations.

You can find the solutions (or **roots**) to a quadratic equation by drawing a graph.

You solve quadratic equations such as $x^2 - x - 4 = 0$ by reading off the x coordinates where the graph $y = x^2 - x - 4$ crosses the x-axis ($y = 0$).

You may also be asked to 'write down the roots' of the equation. This is the same as solving it.

WORKIT!

Here is the graph of $y = x^2 - x - 4$.

Use the graph to solve the equation

$x^2 - x - 4 = 0$

$x = 2.6$ and $x = -1.6$

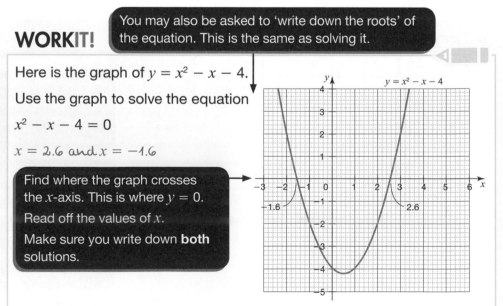

Find where the graph crosses the x-axis. This is where $y = 0$.

Read off the values of x.

Make sure you write down **both** solutions.

You solve quadratic equations such as $x^2 - 3x - 2 = 3$ by reading off the x coordinate at the points of intersection of the graph $y = x^2 - 3x - 2$ with the graph $y = 3$.

DOIT!

On graph paper, draw the graph of $y = x^2$ for values of x between -3 and $+3$. Use your graph to find the approximate values of x when $y = 3$.

WORKIT!

Here is the graph of $y = x^2 - 3x - 2$.

Use the graph to solve the equation $x^2 - 3x - 2 = -3$.

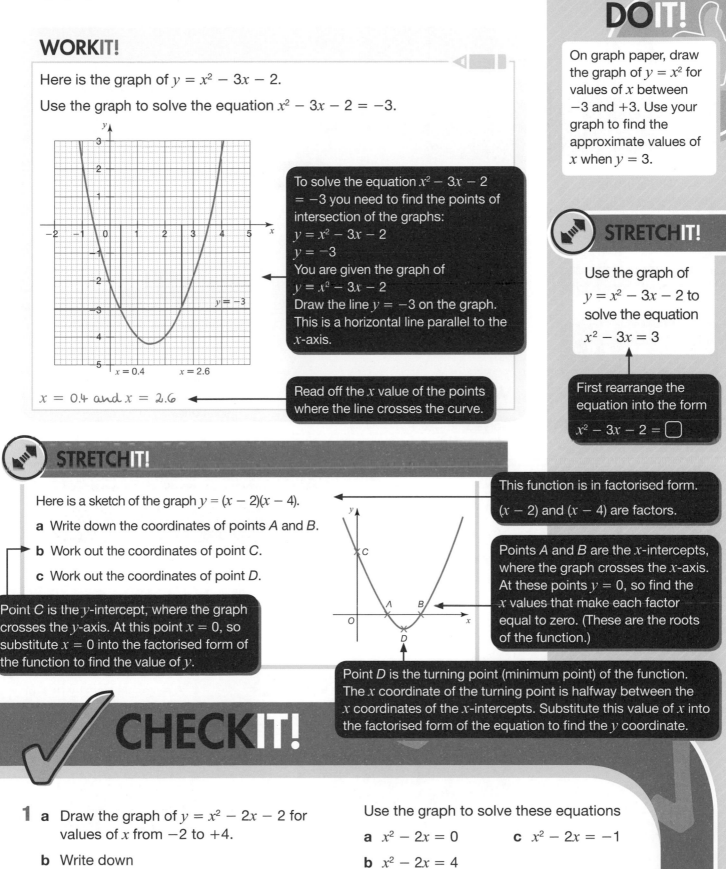

$x = 0.4$ and $x = 2.6$

To solve the equation $x^2 - 3x - 2 = -3$ you need to find the points of intersection of the graphs:
$y = x^2 - 3x - 2$
$y = -3$
You are given the graph of
$y = x^2 - 3x - 2$
Draw the line $y = -3$ on the graph. This is a horizontal line parallel to the x-axis.

Read off the x value of the points where the line crosses the curve.

STRETCHIT!

Use the graph of $y = x^2 - 3x - 2$ to solve the equation
$x^2 - 3x = 3$

First rearrange the equation into the form
$x^2 - 3x - 2 = \square$

STRETCHIT!

Here is a sketch of the graph $y = (x - 2)(x - 4)$.

a Write down the coordinates of points A and B.

b Work out the coordinates of point C.

c Work out the coordinates of point D.

Point C is the y-intercept, where the graph crosses the y-axis. At this point $x = 0$, so substitute $x = 0$ into the factorised form of the function to find the value of y.

This function is in factorised form.
$(x - 2)$ and $(x - 4)$ are factors.

Points A and B are the x-intercepts, where the graph crosses the x-axis. At these points $y = 0$, so find the x values that make each factor equal to zero. (These are the roots of the function.)

Point D is the turning point (minimum point) of the function. The x coordinate of the turning point is halfway between the x coordinates of the x-intercepts. Substitute this value of x into the factorised form of the equation to find the y coordinate.

CHECKIT!

1 a Draw the graph of $y = x^2 - 2x - 2$ for values of x from -2 to $+4$.

   b Write down

      i the equation of the line of symmetry

      ii the coordinates of the turning point.

2 Draw the graph of $y = x^2 - 2x$ for values of x from -2 to $+4$.

Use the graph to solve these equations

a $x^2 - 2x = 0$ c $x^2 - 2x = -1$

b $x^2 - 2x = 4$

3 a Draw the graph of $y = x^2 + 2x - 1$ for values of x from -3 to $+3$.

   b Using your graph, write down the roots of $x^2 + 2x - 1 = 0$.

Solving quadratic equations

A **quadratic equation** contains an x^2 term but no higher power of x. $x^2 - 9 = 0$, $x^2 + 2x = 0$, $x^2 + 4x + 4 = 0$ and $x^2 - 3x - 2 = 0$ are all examples of quadratic equations.

You can solve quadratic equations by factorising.

> If you're unsure about factorising, have a look at page 45.

$$x^2 + 2x = 0$$

$$x(x + 2) = 0 \quad \longleftarrow \boxed{\text{Factorise}}$$

Either $x = 0$ or $x + 2 = 0$.

> Solve the linear equation $x + 2 = 0$

So $x = 0$ or $x = -2$.

To solve an equation of the form $x^2 - 9 = 0$ use the difference of two squares: $x^2 - a^2 = (x + a)(x - a)$.

> $x^2 - 9 = x^2 - 3^2$

$$x^2 - 9 = 0$$

$$(x + 3)(x - 3) = 0$$

Either $x + 3 = 0$ or $x - 3 = 0$.

> Solve the linear equations $x + 3 = 0$ and $x - 3 = 0$

So $x = -3$ or $x = 3$.

DO IT!

Solve the equation $x^2 - 9 = 0$ using this alternative method.
1 Rearrange to make x^2 the subject.
2 Take the square root of both sides.
3 Give the negative and positive solutions.

WORKIT!

Solve $x^2 + 4x + 4 = 0$

$$x^2 + 4x + 4 = 0$$

$$(x + 2)(x + 2) = 0 \quad \longleftarrow$$

> Factorise. The factor pairs of 4 are 1×4 and 2×2. The pair that add up to 4 are 2 and 2.

Either $x + 2 = 0$ or $x + 2 = 0$

So $x = -2 \quad \longleftarrow$

> Solve the linear equation $x + 2 = 0$
> Note that here there is only **one** solution.

Check: $(-2)^2 + 4 \times (-2) + 4 = 4 - 8 + 4 = 0$ ✓

STRETCH IT!

Solve these quadratic equations

$\dfrac{x^2}{2} = 8$

$2x^2 = 50$

$y = x^2 - 4x$ and $y = x^2 - 49$ are both examples of quadratic functions. The roots of a quadratic function are the solutions when $y = 0$.

$$y = x^2 - 4x$$

> Set $y = 0$.

$$x^2 - 4x = 0$$

$$x(x - 4) = 0 \quad \longleftarrow \boxed{\text{Factorise.}}$$

Either $x = 0$ or $x - 4 = 0$.

> Solve the linear equation $x - 4 = 0$.

So the roots of the function are $x = 0$ or $x = 4$.

STRETCH IT!

A rectangle x cm by $(x + 6)$ cm has an area of 40 cm². Find the value of x.

> First write an expression for the area of the rectangle.

CHECK IT!

1 Solve

  a $x^2 - 4x = 0$

  b $x^2 + 7x = 0$

  c $x^2 - 16 = 0$

  d $x^2 + 10x + 9 = 0$

  e $x^2 + x - 12 = 0$

  f $x^2 - 6x - 16 = 0$

2 Find the roots of these functions.

  a $y = x^2 - 49$

  b $y = x^2 - 3x$

  c $y = x^2 + 7x + 6$

Cubic and reciprocal graphs

Cubic graphs

A **cubic function** has an x^3 term and no higher power of x.

$y = x^3$, $y = x^3 + 1$ and $y = x^3 - x$ are all examples of cubic functions.

SNAPIT!   General shape of a cubic graph

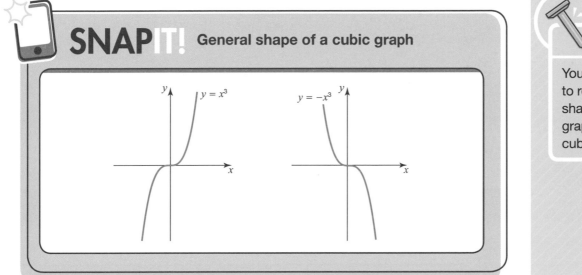

NAILIT!

You need to be able to recognise the shapes of quadratic graphs (page 83) and cubic graphs.

To draw a cubic graph you need to complete a table of values.

WORKIT!

a Complete the table of values for $y = x^3 + 2$

x	-2	-1	0	1
y	-6	1	2	3

Substitute each value of x into $y = x^3 + 2$ to work out the corresponding value of y.

Take care with negative numbers: $(-1)^3 = -1 \times -1 \times -1 = -1$

When $x = -2$, $y = (-2)^3 + 2$
$= -8 + 2 = -6$

b Draw the graph of $y = x^3 + 2$

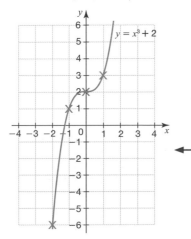

Plot the points from your table on the graph: $(-2, -6)$...

Draw a smooth curve through **all** the points.

DOIT!

Draw the graph of $y = x^3 - 2x$ for values of x between -2 and $+2$.

x	-2	-1
x^3		
$-2x$		
y		

You could use a table of values like this. Make sure you add enough columns.

Reciprocal graphs

The **reciprocal** of x is $\frac{1}{x}$.

$y = \frac{1}{x}$ is an example of a **reciprocal function**.

The graph of a reciprocal function has two parts. It does not cross or touch the x-axis or y-axis.

This is the graph of $y = \frac{1}{x}$.

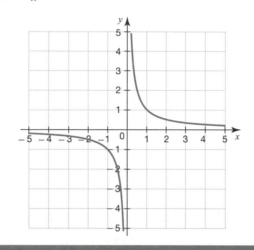

DO IT!

Compare the graph of $y = \frac{1}{x}$ with the graph of $y = -\frac{1}{x}$ (see Check it!). How does the negative sign affect the shape of the graph?

STRETCHIT!

a Complete the table of values for $y = x + \frac{1}{x}$

x	-3	-2	-1	$-\frac{1}{2}$	$-\frac{1}{4}$	$\frac{1}{4}$	$\frac{1}{2}$	1	2	3
y			-2		$-4\frac{1}{4}$					$3\frac{1}{3}$

b Draw the graph of $y = x + \frac{1}{x}$

CHECKIT!

1 a Complete the table of values for $y = x^3 - 1$

x	-3	-2	-1	0	1	2	3
y		-9					26

b Draw the graph of $y = x^3 - 1$

2 a Complete the table of values for $y = -\frac{1}{x}$

x	-4	-3	-2	-1	$-\frac{1}{2}$	$-\frac{1}{4}$	$\frac{1}{4}$	$\frac{1}{2}$	1	2	3	4
y	$\frac{1}{4}$	$\frac{1}{3}$			2		-4			$-\frac{1}{2}$		

b Draw the graph of $y = -\frac{1}{x}$

Drawing and interpreting real-life graphs

DO IT!

Use the information in the table to draw a conversion graph for temperatures in degrees Fahrenheit (°F) and degrees Celsius (°C).

°F	−22	32	140
°C	−30	0	60

Use the graph to convert 40°C to °F.

Linear graphs

Conversion graphs can be used to convert one unit into another. They can also be used to convert between different currencies.

Straight-line graphs used to represent a real-life situation such as a fuel bill will not go through the origin (0, 0) if there is a fixed or standing charge.

WORKIT!

The graph shows the charge for electricity, in £s.

The cost consists of a standing charge plus a charge per unit of electricity used.

a Write down the standing charge.

Standing charge = £20

b Work out the cost per unit of electricity.

Gradient = $\frac{30}{120}$ = 0.25

Cost per unit of electricity is 25p.

> Read off the cost of 0 units from the graph.

> The gradient $\frac{\text{cost (£)}}{\text{number of units used}}$ of the graph gives the cost per unit of electricity.
>
> Draw a large right-angled triangle on the line. Make sure your base is a whole number of units.
>
> Use the scales on the axes to work out the differences in the x and y coordinates.

For some questions you cannot simply read off a value directly from the graph.

WORKIT!

This graph can be used to convert between kilometres and miles.

Use the graph to convert 80 km into miles.

From the graph, 40 km = 25 miles

So 80 km = 50 miles

> The graph does not go up to 80 km.
> Read off the number of miles in 40 km and multiply by 2.

> You could read off the value for 10 km and multiply by 8, or read off the value for 20 km and multiply by 4. However, the value for 40 km can be read off with more ease.

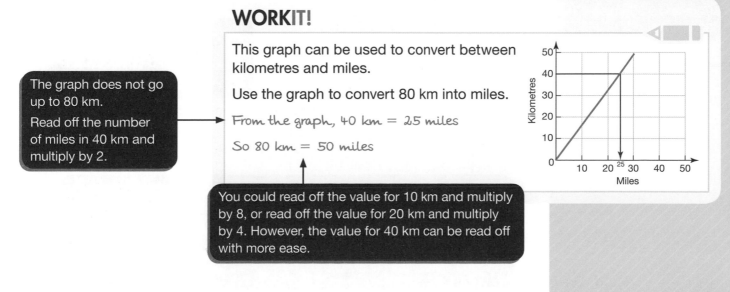

Distance–time graphs

A **distance–time graph** represents a journey. It is an example of a **rate of change** graph, as it shows how distance changes over time.

You can write down information about the journey by looking at the shape of the graph.

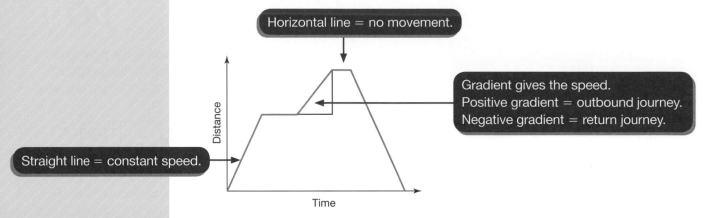

Horizontal line = no movement.

Gradient gives the speed.
Positive gradient = outbound journey.
Negative gradient = return journey.

Straight line = constant speed.

On a distance–time graph the vertical axis represents the distance from the starting point. The horizontal axis represents time. The time could represent 'time taken' given in hours, minutes or seconds, or it could represent the 'time of day' given as, for example, 09:00, 14:00 (using the 24–hour clock).

WORKIT!

Bradley went for a bike ride.

He left home at 7 am.

The distance–time graph represents his bike ride.

At 9 am Bradley stopped for a rest.

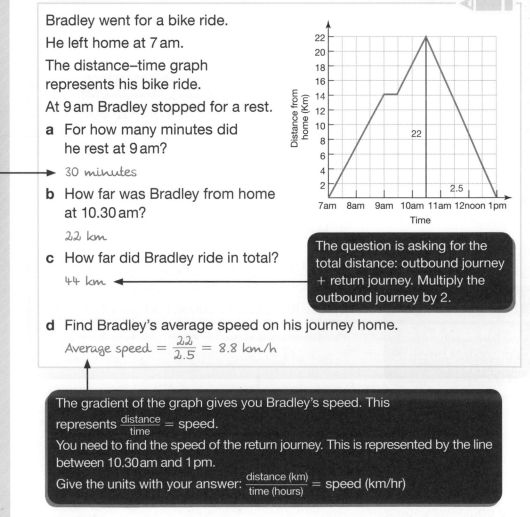

You are told that Bradley stops for a rest at 9 am. The horizontal line represents no movement. He rested between 9 am and 9.30 am so 30 minutes.

a For how many minutes did he rest at 9 am?

   30 minutes

b How far was Bradley from home at 10.30 am?

   22 km

c How far did Bradley ride in total?

   44 km

The question is asking for the total distance: outbound journey + return journey. Multiply the outbound journey by 2.

d Find Bradley's average speed on his journey home.

   Average speed = $\frac{22}{2.5}$ = 8.8 km/h

The gradient of the graph gives you Bradley's speed. This represents $\frac{\text{distance}}{\text{time}}$ = speed.
You need to find the speed of the return journey. This is represented by the line between 10.30 am and 1 pm.
Give the units with your answer: $\frac{\text{distance (km)}}{\text{time (hours)}}$ = speed (km/hr)

Speed–time graphs

A **speed–time graph** also represents a journey. It shows how speed changes over time.

You can write down information about the journey by looking at the shape of the graph.

On a speed–time graph the vertical axis represents speed. The horizontal axis represents time.

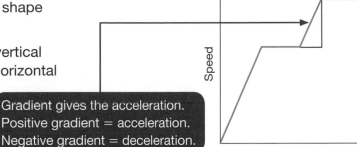

Horizontal line = constant speed (no acceleration).

Gradient gives the acceleration.
Positive gradient = acceleration.
Negative gradient = deceleration.

The gradient of the graph gives you the train's acceleration. This represents

$$\frac{\text{change in speed}}{\text{time}} = \text{acceleration}.$$

The first part of the journey is between 0 and 40 seconds. Give the units with your answer:

$$\frac{\text{change in speed (m/s)}}{\text{time (s)}} = \text{acceleration (m/s}^2)$$

The acceleration is negative. This means that the train is slowing down (decelerating)

WORKIT!

The graph shows information about the journey of a train between two stations.

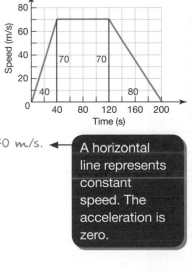

a What is the train's maximum speed?

70 m/s ⟵ Read off the value from the graph.

b Describe the train's journey between 40 and 120 seconds.

The train is travelling at a constant speed of 70 m/s. ⟵

A horizontal line represents constant speed. The acceleration is zero.

c Work out the train's acceleration for the first part of the journey.

Acceleration $= \frac{70}{40} = 1.75$ m/s^2

d Work out the train's acceleration for the final part of the journey.

Acceleration $= -\frac{70}{80} = -0.875$ m/s^2

Interpreting graphs

For a container being filled with water, a graph can be drawn to show how the depth of water changes over time. This is also an example of a rate of change graph.

For a container with straight vertical sides, the graph will be a straight line as it fills at a constant rate. A narrow container will fill faster than a wider container. On a graph this would be represented by a steeper line.

A container with a slanting side will produce a curved graph. For a container with a narrower bottom, the water level rises quickly at first but then slows down as the container widens.

DO IT!

Water is poured into these containers.

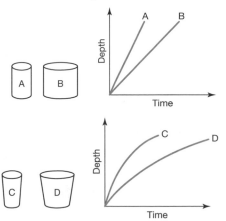

Sketch a graph to show depth against time for each container.

CHECK IT!

1 You can use the information in the table to convert between pounds (£) and euros (€).

Pounds (£)	0	20	50	80
Euros (€)	0	24	60	96

a Use the information in the table to draw a conversion graph.

b Explain how your graph shows that the number of pounds is directly proportional to the number of euros.

c Use the graph to convert

  i £30 to €

  ii €90 to £

Hannah wants to buy a ring.

The ring costs €100 in France.

The same ring costs £90 in the United Kingdom.

d In which country is the ring cheaper? Give a reason for your answer.

2 This graph shows the cost of using a mobile phone for one month.

The cost includes a fixed monthly charge and charge per minute of call above the limit of the contract.

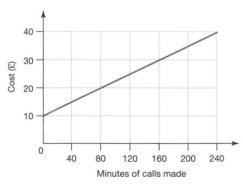

a Write down the monthly charge.

b Work out the charge per minute of calls.

3 The cost of hiring a floor sander is given by the formula $C = 12d + 20$, where d is the number of days for which the sander is hired and C (£) is the total cost of hire.

a Draw the graph of $C = 12d + 20$.

b What is the deposit required for hiring the sander?

Another shop hires out floor sanders where the cost of hire is given by the formula $C = 10d + 30$.

c Which shop would you use to hire the sander for more than 5 days? You **must** show your workings.

4 Elinor drives from her house to her friend's house.

The graph represents part of her journey.

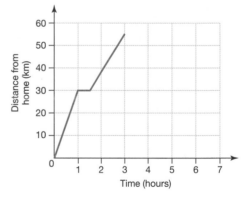

One hour after setting off she stopped for a break.

a For how many minutes did she rest?

Elinor arrived at her friend's house $1\frac{1}{2}$ hours after her break.

b How far is Elinor's house from her friend's house?

c Work out Elinor's speed before and after her break.

Elinor stayed at her friend's house for 2 hours. She then drove home at a steady speed.

It took her 2 hours to get home.

d Complete the graph.

5 A ball is thrown vertically upwards.

The graph shows the path followed by the ball.

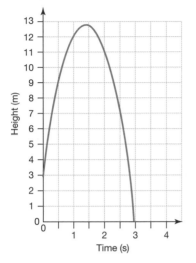

a What is the maximum height reached by the ball?

b How long was the ball in the air?

c There were two times when the ball was at a height of 8 m.

Write down these two times.

d Give a practical reason as to why the graph starts at 3 m.

6 The graph shows information about part of the journey of a lorry between two sets of traffic lights.

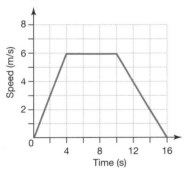

a What was the lorry's maximum speed?

b How long did it take the lorry to reach the maximum speed?

c For how long was the lorry travelling at a constant speed?

d Work out the acceleration of the lorry for the first part of the journey.

7 Part of a cyclist's training session is described below.

A cyclist sets off from a resting position and accelerates to a speed of 10 m/s in 20 s.

She then cycles at a constant speed of 10 m/s for a further 50 s.

She then accelerates to a speed of 20 m/s in 30 s.

a Complete the speed–time graph to show the cyclist's session.

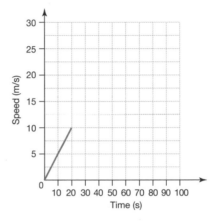

b Work out the cyclist's acceleration over the final 30 s.

8 The graph shows the depth of water (cm) in a bath.

Between *O* (0, 0) and *A* the hot and cold water taps are on.

Between *B* and *C* a person got into the bath.

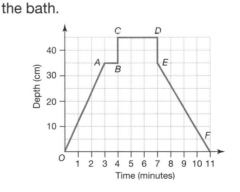

a What was the maximum depth of water before the person got in?

b What was happening between *C* and *D*?

c What was happening between *D* and *E*?

d Which was quicker, running water into the bath or emptying it? Give a reason for your answer.

You may **not** use a calculator for these questions.

1

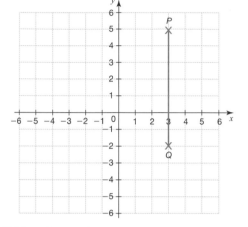

a Write down the coordinates of the point P.

b On the grid, mark with a cross (x) the point $(-4, -2)$. Label this point R.

c On the grid, mark with a cross (x) a point S so that the quadrilateral $PQRS$ is a square.

d Find the coordinates of the midpoint of QS.

e On the grid, draw the line with equation $y = -4$.

2 a Solve $2x + 8 = 4$.

b Here are three expressions.

A $\dfrac{y}{x}$ B $x - y$ C xy

When $x = 2$ and $y = -4$, which expression has the smallest value?

c George says 'When $x = 4$, the value of $3x^2$ is 144'.

Millie says 'When $x = 4$, the value of $3x^2$ is 48'.

Who is correct? Give a reason for your answer.

3 a Simplify $7a - (3a + 4)$.

b Factorise $8x + 12$.

c Simplify $m^4 \times m$.

d Simplify $\dfrac{x^8}{x^3}$

4 a Complete the table of values for $y = 2x + 1$.

x	-1	0	1	2	3
y			3		

b Draw the graph of $y = 2x + 1$.

c Write down the gradient of the line.

5 Solve $4x + 4 = x + 13$.

6 a Write down the inequality shown on the number line.

b Show the inequality $-1 < x \leq 3$ on a number line.

c Write down all the integer values of x that satisfy $1 \leq x < 4$.

d Find the largest integer value of x that satisfies the inequality

$4x + 2 \leq 2x + 5$

7 Luke is x years old.

His grandfather is 65 years older.

His grandfather is also 6 times as old as Luke.

a Write an expression, in terms of x, for his grandfather's age.

b Form an equation in x to work out Luke's age.

8 Here are the first five terms of a sequence.

3, 9, 15, 21, 27

a Write down the next term of this sequence.

b Can 44 be a term in this sequence? Give a reason for your answer.

The nth term of a different sequence is $2n^2 - 3$.

c Work out the 5th term of this sequence.

9 Expand and simplify $(x + 3)(x + 4)$.

10 a and b are whole numbers.

$a > 30$ $b < 20$

Work out the smallest possible value of $a - b$.

11 Here is a rectangle.

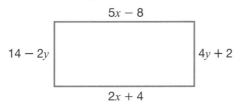

5x − 8

14 − 2y

4y + 2

2x + 4

Work out the value of x and the value of y.

12 a Factorise $4x^2 + 6x$.

b Factorise $x^2 - 100$.

c Solve by factorising $x^2 + 9x + 18 = 0$.

13 Work out the values of a and b in the identity

$3(ax - 4) + 2(4x + b) \equiv 14x - 6$

14 a Simplify $9m - 4m + 7m$.

b Simplify $3p \times 4p$.

c Simplify $12x \div 2$.

🖩 You **may** use a calculator for these questions.

15 a Solve $5(w - 4) = 35$.

b Work out the value of $5a + 7b$ when $a = 7$ and $b = -2$.

c Write an expression for the total cost of 5 cakes at a pence each and 8 biscuits at b pence each.

16 Square *LMNO* has sides of length 6 units. Find the coordinates of point *N*.

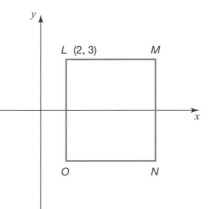

17 Three vertices (corners) of a rectangle *PQRS* are *P* (3, 4), *Q* (3, −2) and *R* (−5, −2).

a Work out the coordinates of vertex *S*.

b Write down the length and width of the rectangle.

18 You can use the information in the table to convert between inches and centimetres.

Inches	0	2	8	12
Centimetres	0	5	20	30

a Use the information in the table to draw a conversion graph.

b Use the graph to convert

i 10 inches to centimetres

ii 50 centimetres to inches.

Arwel needs 60 inches of wooden beading.

The beading costs 2p per centimetre.

c Work out the cost of the beading Arwel needs.

19 Here is a formula.

$4b = a - 8k$

Find the value of k when $a = 80$ and $b = 15$.

20 This rule can be used to work out the cost, in pounds, of hiring a fancy dress costume.

Multiply the number of days by 12.50 and then add 10.

Richard hires a costume for x days.

The total cost is £T.

a Write down a formula for T in terms of x.

Suzanne hires a costume.

The total cost is £72.50.

b For how many days did Suzanne hire the costume?

21 Line L is shown.

Work out the equation of line L.

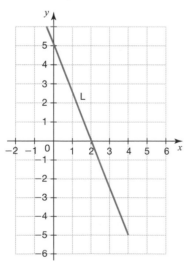

22 a Solve the inequality $4x + 2 \leq 8$.

b Find the integer values of x that satisfy both the inequalities

$3x - 4 < 17$ and $4x + 2 \geq 22$

23 Ollie says $(x + 4)^2 = x^2 + 16$.

Ollie is wrong. Explain his mistake.

24 Make Q the subject of the formula $P = \frac{Q}{4} + R$.

25 a Factorise $m^2 + 8m$.

b Factorise $x^2 + 7x + 12$.

26 Here are the first five terms of an arithmetic sequence.

2, 5, 8, 11, 14

a Find an expression, in terms of n, for the nth term of this sequence.

The nth term of a different sequence is $2n - 3$.

b Kadeena says that 112 is a term in the sequence.

Is Kadeena correct? Give a reason for your answer.

27 a Expand and simplify $4(x + 5) - 3(2x - 1)$.

b Simplify $4a^3b^2 \times 5a^2b$.

28 Here is a quadrilateral.

All measurements are in centimetres.

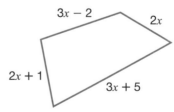

The perimeter of the quadrilateral is 49 cm. Work out the value of x.

29 Here are two number machines, A and B.

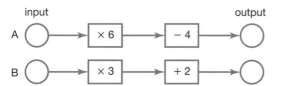

Both machines have the same input.

Work out the input that makes the output of A four times the output of B.

30 An arithmetic progression starts $4 + 2a$, $4 + 4a$, $4 + 6a$.

If the fifth term is 64, work out the value of a.

31 Here are the first three terms of a Fibonacci-type sequence.

$a, b, a + b$

a Show that the 7th term of this sequence is $5a + 8b$.

The 3rd term of the sequence is 5. The 7th term of the sequence is 34.

b Work out the value of a and the value of b.

32 Which equation has roots 2 and -4?

A $x^2 - 8 = 0$

B $(x + 2)(x - 4) = 0$

C $(x - 2)(x + 4) = 0$

D $2x(x - 4) = 0$

Ratio, proportion and rates of change

Units of measure

You need to be able to convert between units of measure.

SNAPIT! Converting between units of measure

Time:	1 year	= 365 days (366 days in a leap year)	Length:	1 km = 1000 m 1 m = 100 cm 1 cm = 10 mm
	1 year	= 12 months	Area:	$1 m^2$ = 10 000 cm^2 $1 cm^2$ = 100 mm^2
	1 day	= 24 hours	Volume:	1 litre = 1000 ml
	1 hour	= 60 minutes		1 ml = 1 cm^3
	1 minute	= 60 seconds	Weight:	1 kg = 1000 g

If you know these standard conversions, you can multiply or divide to convert between units.

NAILIT!

Multiply to convert to a smaller unit of measure.

Divide to convert to a larger unit of measure.

Use your common sense – are you expecting the answer to be smaller or larger after conversion?

1 kg = 1000 g
Since you are moving to a larger unit of measure, you must divide.

Make sure you give the correct unit in the answer.
You could convert all the units to mm instead of cm. If it doesn't say in the question which units to use, you can use either.

WORKIT!

1 Convert 2.4 km into m.

$2.4 \times 1000 = 2400$ m

1 km = 1000 m
Since you are moving to a smaller unit of measure, you must multiply.

2 Convert 720 g into kg.

$720 \div 1000 = 0.72$ kg

WORKIT!

1 How many minutes are there in half a day?

Half a day = 12 hours

$12 \times 60 = 720$ minutes

Multiply, since the unit is smaller.

2 Work out the perimeter of this rectangle.

$125 \div 10 = 12.5$ cm

Perimeter = $12.5 + 12.5 + 3.2 + 3.2$

= 31.4 cm

125 mm

3.2 cm

Divide by 10 since the unit is larger.

Sometimes you will be given the conversion to use. Just use the same rule – multiply if you are moving to a smaller unit of measure, divide if you are moving to a larger unit of measure.

WORKIT!

A length of wood measures 17 inches. What is its approximate length in centimetres?

1 inch ≈ 2.5 cm

$17 \times 2.5 \approx 42.5\,cm$

Multiply since the units are smaller.

NAILIT!

≈ means 'is approximately equal to' and is used when the conversion isn't exact.

DOIT!

Work out:

a how many seconds you spent doing maths work today

b how long your arm is in metres

c how much you weigh in grams

d the area in m² of a square with sides measuring 10 cm by 10 cm.

SNAPIT!

Ratios of lengths, areas and volumes

Line A is a cm and Line B is b cm so their lengths are in the ratio $a : b$

A ____   B ____

Square A is $a \times a = a^2$ cm² and square B is $b \times b = b^2$ cm² so their areas are in the ratio $a^2 : b^2$

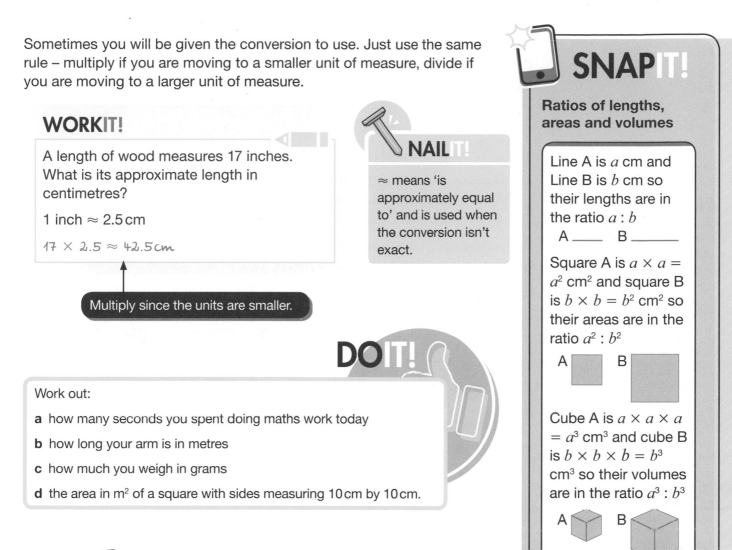

A   B

Cube A is $a \times a \times a = a^3$ cm³ and cube B is $b \times b \times b = b^3$ cm³ so their volumes are in the ratio $a^3 : b^3$

A   B

✓ CHECKIT!

1 Convert these measurements to the units shown in brackets:

a 3 km (m)

b $1\frac{1}{4}$ hours (mins)

c 1.3 m² (cm²)

d 3520 ml (litres)

e 2 hours (seconds)

f 14 000 g (kg)

2 A clothed baby weighs 4.5 kg. If his clothes weigh 325 g, what is the actual weight of the baby?

3 A bag weighs 5 pounds. What is the weight in kg?

1 kg ≈ 2.2 pounds

Ratio

A ratio is a way of expressing the relationship between two different quantities.

For example, the ratio of the number of girls to the number of boys in a class is **5:4**. This tells you that for every **5** girls there are **4** boys. If there are 10 girls (2 × **5**) we know there will be 8 boys (2 × **4**).

You can cancel a ratio down by dividing all the numbers by the same amount. You can find equivalent ratios by multiplying all the numbers by the same amount.

For example, the ratio 5:4 is equivalent to the ratio 10:8 or 15:12 or 2.5:2.

A ratio can be reduced to its simplest form by dividing all the parts by the highest common factor. ◄── For more on the highest common factor, see page 10.

WORKIT!

Simplify the ratio 4:8:32.

1:2:8 ◄── The highest common factor of 4, 8 and 32 is 4, so divide each part of the ratio by 4.

It may be useful to write a ratio in the form 1:n.

To mix deep purple paint an artist uses 50 ml of blue paint and 125 ml of red paint.

The ratio of blue to red paint is 50:125.

Dividing both parts of the ratio by 50 gives: 1:2.5.

The relationship between the volumes of blue and red paint can be described as:

'1 part blue to 2.5 parts red'.

Expressing a relationship as a ratio

If one number or quantity is a multiple of another, you can express the relationship as a ratio.

If the number of girls in the class is three times the number of boys then you can say that for every 3 girls there is 1 boy, and that the ratio of girls to boys is 3:1.

A ratio can be simplified into the form 1:n by dividing.

For example to write the ratio 2:3 in the form 1:n divide all the parts of the ratio by 2 giving: 1:$\frac{3}{2}$ or 1:1.5.

WORKIT!

A bag contains red and blue balls. $\frac{5}{8}$ of the balls are red.

Write down the ratio of red balls to blue balls.

The ratio of red to blue balls is 5:3 ◄── If $\frac{5}{8}$ of the balls are red and the rest are blue then $\frac{3}{8}$ must be blue.

Sharing in a given ratio

To share a quantity in a given ratio, split it into the number of 'parts' given by the ratio.

If Amy and Rudi share £90 in the ratio 4:5, think of the parts of the ratio as boxes to fill.

Amy				Rudi				

There are $4 + 5 = 9$ boxes to fill.

Each box must contain the same number.

£10	£10	£10	£10	£10	£10	£10	£10	£10
Amy				Rudi				

$90 \div 9 = £10$

Amy receives $4 \times £10 = £40$.

Rudi receives $5 \times £10 = £50$.

WORKIT!

A DIY store mixes 'Woodland Green' paint using the colours blue, yellow and white in this ratio:

1:3:20

If the amount of white base used for the paint is 2 litres, how much green paint will the customer get?

Total amount of paint is $1 + 3 + 20$ parts $= 24$ parts.

2 litres $= 20$ parts

1 part $= \frac{2}{20} = \frac{1}{10}$ litre $= 100$ ml

Total paint $= 100 \times 24 = 2400$ ml $= 2.4$ litres

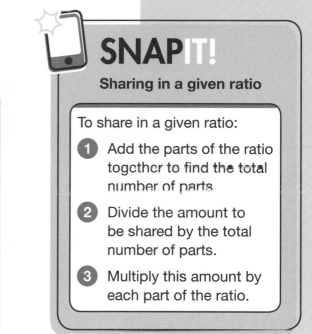

SNAPIT!

Sharing in a given ratio

To share in a given ratio:

1. Add the parts of the ratio together to find the total number of parts

2. Divide the amount to be shared by the total number of parts.

3. Multiply this amount by each part of the ratio.

WORKIT!

To make a soft drink, cordial and water are mixed in the ratio 2:15.

A bottle contains 1 litre of cordial.

a How many litres of water are needed to mix the drink?

$15 \div 2 = 7.5$ litres ◄— The ratio 2:15 means that for every 2 litres of cordial you need 15 litres of water.
If you only have 1 litre of cordial, divide by 2 to work out the volume of water required.

b How many litres of the drink will be made?

$1 + 7.5 = 8.5$ litres ◄— Add together the volume of the cordial and the water.

c How many litres of cordial will you need to mix 170 litres of the drink?

$2 + 15 = 17$

$170 \div 17 = 10$ ◄— Share 170 in the ratio 2:15 using the steps in the Snap it! box above.

Amount of cordial $= 2 \times 10 = 20$ litres

STRETCHIT!

Sines, cosines and tangents are special ratios – the ratios between the sides of a right-angled triangle. For more on trigonometric ratios see page 160.

Writing ratios as functions

If a ratio is in the form $1:a$, it can be written as a function.

For example, if the number of girls to boys is $1:2$ then for every 1 girl there are 2 boys.

You can write this as the equation $b = 2g$ where $b =$ boys, $g =$ girls.

If there are 4 girls, you could work out the number of boys by substituting $g = 4$ into the function:

$b = 2 \times 4 = 8$ boys

WORKIT!

A recipe asks for the number of eggs to the amount of sugar, in grams, in the ratio $1:120$.

Write down a function relating the number of eggs (e) to the amount of sugar (g).

$g = 120e$ ← The grams of sugar needed are 120 times the number of eggs.

✓ CHECKIT!

1 Simplify each ratio.

a $3:12$ c $8:10$

b $5:15:20$

2 On a school trip there are 35 students and 5 teachers.

Write down the ratio of students to teachers. Simplify your answer.

3 A mix for concrete requires 250 g of cement to 375 g of sand. Copy and complete the statement: 'allow 1 part cement for … parts sand'.

4 A cinema sells 7 times more tickets in the evening than in the afternoon.

a Write down the ratio of the number of tickets sold in the evening to the number of tickets sold in the afternoon.

b In one day, 800 tickets were sold. How many were sold in the afternoon?

5 A survey suggests that $\frac{3}{5}$ of cats prefer Pawpaw cat food.

a What is the ratio of cats that prefer Pawpaw cat food to those who don't?

b The survey tested 200 cats. How many preferred Pawpaw cat food?

6 The ratio of sides in a triangle is $3:4:5$.

If the shortest side has a length of 9 cm what are the lengths of the two longer sides?

7 The ratio of teachers to students in a school is $1:20$.

Write down a function relating the number of teachers (t) to students (s).

8 In a recipe the total volume of flour and sugar required is 1.5 kg.

a If the ratio of sugar to flour is $2:3$, how much sugar is needed?

A different recipe asks for sugar and flour in the ratio $2:5$.

b If there is 60 g more flour than sugar, how much sugar is there in this recipe?

Scale diagrams and maps

Scale diagrams are used when you cannot draw things their actual size.

This is a map showing part of Hampshire.

The scale is 1 : 2 000 000.

This means 1 cm on the map represents 2 000 000 cm in real life.

On this map, the distance between Petersfield and Guildford is 1.75 cm.

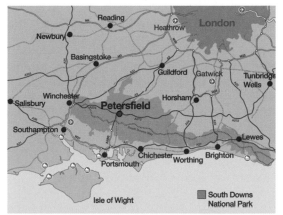

© Petersfield Town Council (contains Ordnance Survey data © Crown copyright and database right 2010)

To work out the distance in real life, multiply by 2 000 000.

Distance = 1.75 × 2 000 000 = 3 500 000 cm

Then convert this into km:

3 500 000 cm = 35 000 m ◄── Divide by 100 since 1 m = 100 cm.

35 000 m = 35 km ◄── Divide by 1000 since 1 km = 1000 m.

NAILIT!

To convert dimensions on a map to real life multiply them by the large number in the scale. Divide real life distances to convert them to map dimensions.

DOIT!

Using the map shown, work out the real-life distances between:

a Portsmouth and Chichester

b Southampton and Portsmouth

c Guildford and Lewes.

WORKIT!

The diagram shows the scale drawing of a ship and the harbour. How far and on what bearing is the ship from the harbour?

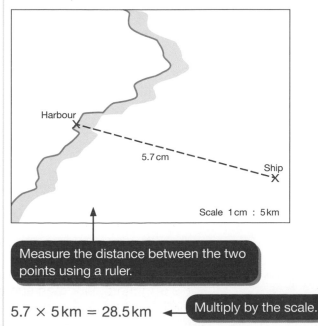

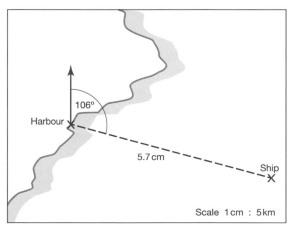

Measure the distance between the two points using a ruler.

Draw a north line at the harbour, then measure the bearing from North. The bearing is 106°.

5.7 × 5 km = 28.5 km ◄── Multiply by the scale.

See page 131 for more on bearings.

If the scale on a map is $1:x$, how would you work out the length in cm of a line representing 50 miles?

Estimating length using a scale

If you know the actual length of something shown in a scale drawing you can use that object to estimate the height of other objects.

In this diagram the beacon is approximately 3.5 times the height of the man.

Therefore, the height of the beacon = $3.5 \times 1.8 = 6.3$ metres.

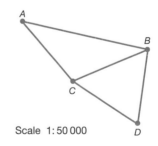

1.8 m

CHECK**IT!**

1 The scale on a map says $1:50000$.

Which of the following is true? You may choose more than one answer.

A   1 cm represents 50 000 cm.

B   1 cm represents 0.5 km.

C   1 cm represents 50 000 m.

D   1 cm represents 5 km.

E   1 cm represents 5000 m.

F   1 cm represents 500 m.

2 A map has a scale 1 cm : 12 km.

   a The distance on the map between town A and town B is 3 cm. What is the distance in real life?

   b The distance between town B and town C is 15 km. How far apart are they on the map?

3 A scale model of a boat is made using a scale of $1:1000$.

If the scale model is 12 cm long, how long is the original boat in metres?

4 The diagram shows an underground railway network.

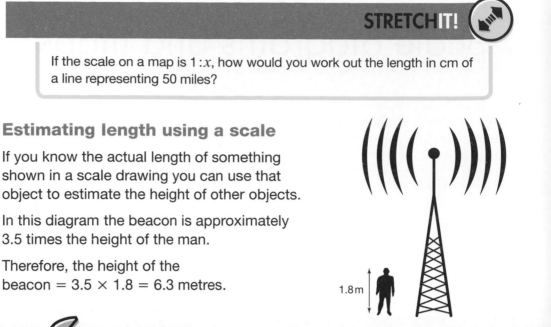

Scale 1 : 50 000

   a What is the distance in km between stations A and C?

   b What is the bearing from B to C?

Fractions, percentages and proportion

If there are 7 men and 3 women in a group, you could express the relationship between the number of men and women in several different ways.

- The **ratio** of men to women is $7:3$.
- The **fraction** of men in the group is $\frac{7}{10}$.
- The **percentage** of men in the group is 70%.

Writing one quantity as a fraction of another

A fraction can be used to express a proportion of the whole, as in the examples below.

The shape has been split into 15 pieces; 6 of these are shaded.

So $\frac{6}{15}$ of the shape has been shaded.

7 can be written as a fraction of 10 by writing $\frac{7}{10}$.

9 cm can be written as a fraction of 1 metre by writing $\frac{9}{100}$. ◄———

36 hours can be written as a fraction of 1 day by writing $\frac{36}{24} = 1\frac{1}{2}$. ◄———

> When writing one quantity as a fraction of another, make sure both quantities are in the same units!

> Sometimes the fraction will be larger than 1.

Writing one quantity as a percentage of another

SNAP IT! Writing one quantity as a percentage of another

1. Write the quantities as a fraction.
2. Convert this to a percentage:

   Multiply the top and bottom by the same amount to give you 100 in the denominator.

DO IT!

Work out what proportion of your day you have spent doing maths work. Express the value as a fraction and as a percentage.

WORKIT!

Alin and Jackie share a prize in the ratio $2:3$.

a What fraction does Jackie receive?

$\frac{3}{5}$ ◄——— Jackie receives 3 parts out of a total of $2 + 3 = 5$ parts.

b What percentage does Alin receive?

$\frac{2}{5}$ ◄——— Alin receives 2 parts out of a total of 5 parts.

$= \frac{40}{100}$ ◄——— Multiply the numerator and denominator by 20 to get 100ths.

$= 40\%$

Proportion

When we talk about proportion we are thinking about what 'part' a number has in relation to a whole.

Proportions can be expressed as a fraction or a percentage.

For example, if there are 10 people in a room and 9 of them are men, the proportion of men is $\frac{9}{10}$ or 90%.

You can compare proportions by considering them as ratios – if the ratios are equal then the proportions are the same.

WORKIT!

> Write both proportions as ratios and compare the ratios. If the ratios are equal then the proportions are equal.

A recipe for cake mix requires 120 g of sugar and 40 g of butter.

Granny says the recipe is wrong and you should use the rule '1 part butter to 3 parts sugar'.

Show that the proportion of sugar to butter is the same in the recipe and using Granny's rule.

> Cancel the ratio down as much as possible.

(Recipe)   butter : sugar = 40 : 120 = 1 : 3

(Granny's rule)   butter : sugar = 1 : 3

Since the ratios are equal, the proportions in both recipes are equal.

✓ CHECKIT!

1 What fraction of 35 m is 20 cm?

2 During a day Alice spends:

   2 hours riding

   3 hours revising for exams

   8 hours sleeping.

   What fraction of the day does she have remaining for other activities?

3 In a box of 20 sweets, 15 are chocolates.

   a What fraction are chocolates?

   b What percentage of the sweets are **not** chocolates?

4 In a recipe, the ratio of butter to sugar to flour is 1 : 2 : 7.

   What percentage of the mixture is butter?

5 The table below shows the number and gender of students at two schools.

	School A	School B
Boys	125	100
Girls	145	120

Is the proportion of girls to boys the same in both schools? Justify your answer.

6 A particular type of biscuit contains 22 g of fat in every 100 g. A 150 g packet of biscuits contains 8 biscuits.

   If you eat one biscuit how many grams of fat do you consume?

Direct proportion

Two things are in direct proportion when they increase at the same rate.

For example, if you are paid £10 an hour, the number of hours you work is in direct proportion to how much you are paid.

If x and y are in direct proportion, an equation linking them is of the form $y = mx$, where m is a numeric value.

Calculations involving direct proportion

To answer questions involving direct proportion, you must decide whether to multiply or divide. Think logically about whether you expect the values to increase or decrease.

There is often more than one way of solving problems involving proportion.

WORKIT!

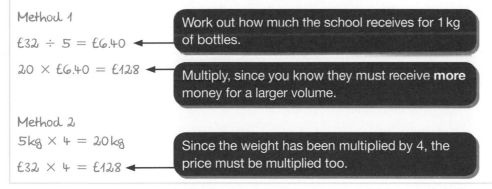

For every 5 kg of plastic bottles a school recycles, they receive £32.

How much will they receive for 20 kg of plastic bottles?

Method 1

£32 ÷ 5 = £6.40   ◄ Work out how much the school receives for 1 kg of bottles.

20 × £6.40 = £128   ◄ Multiply, since you know they must receive **more** money for a larger volume.

Method 2

5 kg × 4 = 20 kg

£32 × 4 = £128   ◄ Since the weight has been multiplied by 4, the price must be multiplied too.

Graphs showing direct proportion

A graph showing two things in direct proportion will be a straight line through the origin.

The equation of the line will always be in the form $y = mx$, where m is a constant.

For example, if you are paid £10 per hour for work then:

total earned = 10 × number of hours

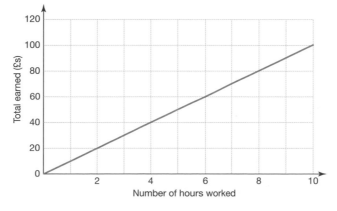

(Graph: vertical axis "Total earned (£s)" from 0 to 120; horizontal axis "Number of hours worked" from 0 to 10; straight line through origin.)

NAILIT!

Equations like $y = 2x$, $a = 3b$ and $m = 0.4n$ are all equations that show direct proportion.

NAILIT!

It doesn't matter which method you use as long as you show your workings.

DOIT!

Decide which of the above methods you prefer for solving this problem.

STRETCHIT!

A taxi driver charges a fixed fee of £2 plus £0.70 per mile. Explain why this is not an example of direct proportion.

Let y = total earned and x = number of hours worked.

The equation of the graph is $y = 10x$.

The gradient of the graph is 10, since for an increase of 1 in the number of hours worked the total earned will increase by £10.

WORKIT!

The graph shows the price of a taxi for different length journeys.

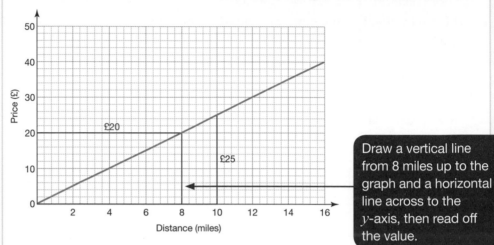

You want to calculate how much the price increases for an increase in distance of 1 mile.

Choose two points on the graph and draw a right-angled triangle, marking on the increase in distance and price.

In this case, choosing the points (0, 0) and (10, 25) means you can easily see that for an increase of 10 miles an extra £25 is charged.

Divide £25 by 10 to find the price increase per mile.

Draw a vertical line from 8 miles up to the graph and a horizontal line across to the y-axis, then read off the value.

a How much does it cost to travel 8 miles?

£20

b How much does each extra mile cost?

£2.50

c Calculate how much a journey of 30 miles would cost.

30 × £2.50 ← Each mile costs £2.50.

= £75

CHECKIT!

1 x and y are in direct proportion. Which of these could be an equation linking them?

A $y = 0.7x$

B $y = x + 0.7$

C $y = x - 0.7$

D $y = \frac{0.7}{x}$

E $y = 7x$

There may be more than one answer.

2 A recipe lists the ingredients for 20 meringues as 2 egg whites and 120 g sugar.

a How many meringues can you make with:

i 3 eggs   **ii** 600 g of sugar?

b How many eggs would you need for 70 meringues?

3 A swimming pool fills at the rate of 4.5 litres per minute. How long will it take before the pool contains 675 litres? Give your answers in hours and minutes.

4 The graph below shows the relationship between the weight of flour and sugar needed for a recipe.

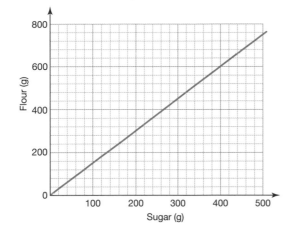

The gradient of the graph is 1.5. Which of the following statements are true?

A If you multiply the weight of sugar by 1.5 you will calculate the weight of flour.

B The weight of sugar is 1.5 times the weight of flour.

C For every 1.5 g of sugar you need 1 g of flour.

D For every 1 g of sugar you need 1.5 g of flour.

Inverse proportion

Two things are in inverse proportion if as one increases the other decreases at the same rate.

For example, as the speed at which you travel goes up, the time your journey will take goes down.

DOIT!

Decide if each of these situations describes things in direct proportion or inverse proportion:

a The number of people painting a fence and the time taken.

b The number of tickets purchased and the total price.

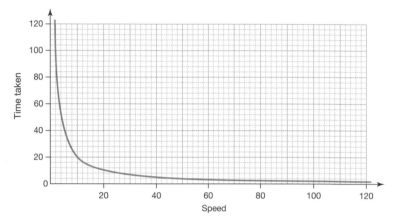

If x and y are in inverse proportion, an equation linking the two is of the form $y = \frac{m}{x}$, where m is a numeric value.

NAILIT!

Equations like $y = \frac{2}{x}$, $a = \frac{3}{b}$ and $m = \frac{0.4}{n}$ are all equations that show inverse proportion. Their graphs all look similar to this:

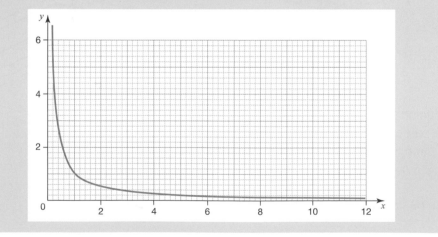

SNAPIT! Proportionality

If y is inversely proportional to x, then y is proportional to $\frac{1}{x}$.

Calculations involving inverse proportion

You should always think carefully about whether to multiply or divide. Are you expecting the value to increase or decrease?

WORKIT!

It takes 5 minutes to milk one cow using a milking machine. A farm purchases 4 milking machines. How long will it take to milk a herd of 100 cows?

Method 1

5 × 100 = 500 minutes ◄— Work out how long it would take 1 machine to milk 100 cows.

500 ÷ 4 = 125 minutes ◄— If there are more machines it will take less time, so divide by 4.

= 2 hours 5 minutes

Method 2

100 ÷ 4 = 25 ◄— They can milk 4 cows at a time, so divide by 4.

25 × 5 = 125 minutes ◄— Multiply by the number of minutes it will take.

= 2 hours 5 minutes

NAILIT!

Use your common sense to answer questions involving direct and inverse proportion.

For example, if you want to boil 100 eggs and each egg needs 3 minutes, you shouldn't boil the eggs for 3 × 100 = 300 minutes = 5 hours!

CHECKIT!

1 Two amounts, x and y are in inverse proportion. Which of these could be an equation that links them?

A $y = 0.7x$

B $y = x + 0.7$

C $y = x - 0.7$

D $y = \frac{0.7}{x}$

E $y = 7x$

2 If Hanisha travels at 60 miles per hour she can get to work in 15 minutes. How long will it take her if she travels at 40 miles per hour?

3 It takes two decorators five days to decorate the outside of a house. How long would it take six decorators?

4 The graph shows the age of chickens on a farm and the average number of eggs they lay every week.

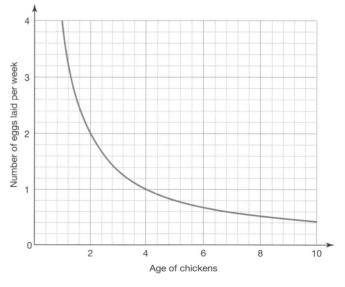

Age of chickens

a How many eggs does a two-year-old chicken lay every week?

b Describe the relationship between the age of the chickens and the number of eggs they lay.

Working with percentages

Percentage change

When things increase or decrease you can calculate the percentage change.

If a price increases by 50% then the new price is 150% of the old price. Alternatively, you could say it is $1\frac{1}{2}$ times the original value or 1.5 times the original value.

> Percentage change can also be expressed as a fraction or a decimal.

WORKIT!

The number of students taking Mandarin at A level in 2017 is 75% of the number who took it in 2016.

Which of the following statements is true? Explain your answers.

a The number of people taking Mandarin has decreased by 25%.

   True – 75% is 25% less than 100%.

b The number of people taking Mandarin in 2017 is 0.25 times the number who took it last year.

   False – the number of people taking Mandarin in 2017 is 0.75 times the number who took it in 2016 since 75% = 0.75.

WORKIT!

The price of juice has risen from 89p per litre to £1.25 per litre. Work out the percentage change. Give your answer to the nearest percentage.

Increase in price = 125 − 89 = 36p

$\frac{36}{89} \times 100 = 40\%$ increase

Comparing two quantities using percentages

We can compare two quantities by considering what percentage one is of the other.

For example, if two different companies sell a DVD at different prices:

Company A = £15   Company B = £12

We could say that Company A charges: $\frac{15}{12} \times 100 = 125\%$ of the price Company B charges.

We could also say that Company B charges: $\frac{12}{15} \times 100 = 80\%$ of the price Company A charges.

Percentage profit or loss

Profit or loss can be calculated in exactly the same way as percentage change.

Calculating original values

If you are given the value after an increase or decrease, you may need to find the original value.

NAIL IT!

When dealing with money a percentage increase is called a profit and a percentage decrease is called a loss.

SNAP IT!

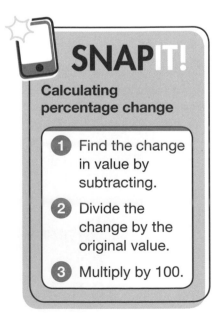

Calculating percentage change

1 Find the change in value by subtracting.

2 Divide the change by the original value.

3 Multiply by 100.

For example, after a 12% increase in price a DVD player costs £56.

112% of the original price = £56

1% of the price = £56 ÷ 112 = £0.50

The original price is 100%:

100 × £0.50 = £50

WORKIT!

In a sale, all prices are reduced by 40%. If the new price of a video game is £18, what was the original price?

60% of the original price = £18

1% of the original price = £18 ÷ 60 = £0.30

100% of the original price = £30.00

SNAPIT! Finding the original price

1. Write down the percentage of the price you are given.

2. Divide the price by the percentage to work out 1%.

3. Multiply by 100 to give the original price.

Calculating interest

When you save money in a bank you will often be paid **interest**, at a stated percentage rate, if you leave the money in the account. This is usually calculated annually (at the end of the year).

To work out how much money will be in the account, calculate:

total increase + original savings

When you only receive interest on the initial investment, this is called **simple interest**.

For example, if you put £100 in a bank with 5% simple interest, then each year you will receive £5.

WORKIT!

A bank offers 2.5% interest for a savings bond. Paul invests £3500. If he does not withdraw any money, how much will he have at the end of the year?

2.5% of £3500 = 0.025 × 3500

= £87.50

£3500 + £87.50 = £3587.50

STRETCHIT!

A credit card charges 28% APR (annual percentage rate). This means if you do not pay off the money owed, the bank will charge an additional 28%.

If you spend £100 how much will you have to pay the credit card company back?

Growth and decay

If something increases year on year, it experiences **growth**.
If something decreases year on year, it experiences **decay**.
For example, the population of the UK increases by approximately 5% each year.

If the population in 2016 is approximately 64 million, you can predict the population over the next few years in the following way:

2017   105% of 64 000 000 = 1.05 × 64 000 000 = 67 200 000
2018   105% of 67 200 000 = 1.05 × 67 200 000 = 70 560 000
2019   105% of 70 560 000 = 1.05 × 70 560 000 = 74 088 000

… and so on.

To work out what the population would be after 10 years, multiply by 1.05 ten times.

A quicker way to do this is to calculate $1.05^{10} \times 64\,000\,000 = 104\,249\,256$.

WORKIT!

The number of bacteria in a petri dish decreases by 10% each hour.

There are 32 000 bacteria in the dish. How many will there be after:

a 1 hour

$0.9 \times 32\,000 = 28\,800$

> If the number of bacteria decreases by 10% there will be 90% left, so find 90% of 32 000.

b 5 hours

$0.9^5 \times 32\,000 = 18\,896$

> You could multiply by 0.9 five times, but it is quicker to multiply by 0.9^5
>
> Round your answer to a sensible degree of accuracy, in this case the nearest whole number.

Compound interest

When you receive interest on your interest in a bank account, this is called **compound interest**.

You calculate compound interest in exactly the same way as any other growth.

For example, if you invest £300 in a bank account with 3% compound interest you could work out how much you would have after 10 years as shown on the right.

> The number of years you have invested the £300 for.

$$£300 \times 1.03^{10} = £403.17$$

> Original investment.

> Multiply by 1.03 since you wish to find 103% of the original investment.

STRETCH**IT!**

You invest £100 in a bank account. What percentage interest would you need to have if you wish to have more than £110 after 5 years?

SNAPIT!   Financial terms

Profit	How much money is made
Loss	How much money is lost
Cost price	The price raw materials cost to purchase
Selling price	The price a product is sold for
Debit	Money that you pay or owe – if your bank balance is £100 in debit it means you owe the bank £100
Credit	Money that you earn, receive or possess – if your bank account has £100 credit it means you have £100 to spend
Income tax	The amount you pay the government when you earn money, worked out according to the amount you earn
VAT	Value Added Tax – the tax the government adds to the price of 'luxury' items; it is currently 20% in the UK
Interest rate	How much money you will be paid if you leave your investments in a bank account or similar; it is usually a percentage of the money you invest, paid annually or over some other time frame

If you borrow money you are the one who has to pay interest – to the bank. The percentage you pay is usually higher than the rate the bank pays to its investors. This is how banks make a profit.

CHECKIT!

1 Work out

   a 103% of £50   **c** 19.5% of 64

   b 248% of 400

2 A music teacher puts his prices up from £40 per hour to £45 per hour. Work out the percentage increase.

3 The average temperature in a Mediterranean country one year was 24°C, 15% greater than in the previous year. What was the average temperature in the previous year?

4 The number of hits on a website increases at a rate of 20% per hour. Initially there are 15 000 hits. How many will there have been after 3 hours?

5 In a sale the price of a TV was reduced by 20%. If it was reduced by £30, what was the original price?

6 In Year 1 a company has 225 employees. In Year 2 there are only 200.

   Copy and complete the statement:

   The number of employees in Year 2 is …% of the number in Year 1.

Compound units

A compound measure is made up of two or more units.

Examples of compound units include speed, density, pressure and rates of pay.

Speed can be measured in, for example, km/h, m/s or mph (miles per hour).

Density can be measured in, for example, g/cm^3 or kg/m^2.

Rates of pay and prices

When working with compound units involving rates of pay and prices, make sure you are working in the same units.

WORKIT!

A call centre charges 13p per minute for calls. How much will a half hour call cost?

13p × 30 mins = 390p ← Convert $\frac{1}{2}$ hour into minutes.

= £3.90

Speed

The relationship between speed, distance and time is illustrated in the triangle.

Cover the unit you wish to calculate:

Speed = distance ÷ time

Distance = speed × time

Time = distance ÷ speed

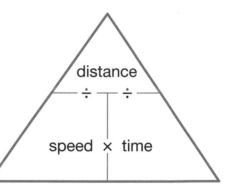

WORKIT!

How far will a tram travelling at 75 km/h travel in 20 minutes?

Distance = speed × time

Speed = 75 km/h

Time = 20 minutes = 1 __ 3 ~~hour~~   Convert the time into hours so the units are the same.

Distance = 75 × $\frac{1}{3}$

= 25 km

If you travel 100 miles in x hours, write an expression to describe your speed in miles per hour.

Density

The relationship between density, mass and volume is illustrated in this triangle.

Density = mass ÷ volume

Mass = density × volume

Volume = mass ÷ density

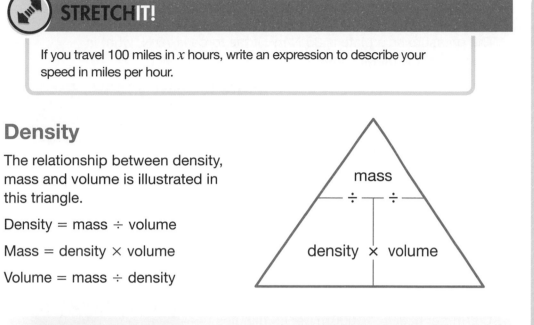

WORKIT!

Work out the density of an object that weighs 50 kg and has a volume of 32 m³. Give your answer to 2 decimal places.

Density = mass ÷ volume

= 50 kg ÷ 32 m³

= 1.56 kg/m³

The unit is kg/m³ since the original units were kg and m³.

Pressure

The relationship between pressure, force and area is illustrated in this triangle.

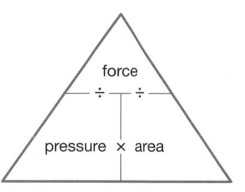

DOIT!

Write down the three formulae that connect pressure, force and area.

DOIT!

Practise drawing out the triangles illustrating the relationships between speed/distance/time, density/mass/volume and pressure/force/area units.

WORKIT!

The pressure on a plate is 35 N/m². If the force is 70 N, what is the area of the metal plate?

Area = force ÷ pressure

   = 70 N ÷ 35 N/m²

   = 2 m²

> Since the force is in Newtons (N) and the pressure is Newtons per m² (N/m²) you do not need to convert units.

Converting compound units

When converting compound units, convert one unit at a time.

For example, to convert 18 km per hour into metres per second:

> There are 1000 m in 1 km.

18 km per hour = 18 000 m per hour

18 000 m per hour = 18 000 m per 60 minutes

> There are 60 minutes in 1 hour.

> Divide by 60 to work out how many m per minute.

= 300 m per minute

300 m per minute = 300 m per 60 seconds

> There are 60 seconds in 1 minute.

> Divide by 60 to work out how many m per second.

= 5 m per second

WORKIT!

A sewing machine sews at 1 inch per second. Work out the speed in metres per minute.

> First convert into cm.

1 inch = 2.5 cm

1 inch per second = 2.5 cm per second

> Work out how many cm it will sew in 1 minute.

2.5 × 60 = 150 cm per minute

= 1.5 m per minute

> Convert into metres.

NAILIT!

A shortcut to convert from km per hour to m per second is to divide by 3.6.

CHECKIT!

1 Electricity costs 18p per unit. If the bill at the end of the month is £29.50, how many units of electricity have been used? Give your answer to the nearest whole number.

2 A train travels at 120 miles per hour. How long will it take to travel 80 miles? Give your answer in minutes.

3 The volume of a cork is 3 cm³. It weighs 0.72 g. What is the density?

4 A force of 12 N is applied to an area of 2 m². Calculate the pressure.

5 Convert 3 m/s into km/h.

6 A petrol pump flows at 0.6 litres per second. Convert the rate of flow into gallons per hour.

   1 gallon = 4.55 litres

7 Usain Bolt ran 100 m in 9.58 seconds. The top speed of a cheetah is 120 km/h.

   Who is faster? Show all your calculations.

⊘ You may **not** use a calculator for these questions.

1 Work out how many

   a m in 3.2 km

   b seconds in 9 minutes

   c ml in 0.4 litres

2 Convert 4600 m into km.

3 How many minutes are there in 2.5 hours?

4 Calculate the area of this shape.

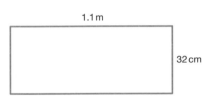

1.1 m

32 cm

5 A square has area 3 m². What is its area in cm²?

6 What fraction of 12 is 5?

7 A bag contains 26 blue balls and 18 red balls. Write the ratio of blue balls to red balls. Give the ratio in its simplest form.

8 Boiling water in a kettle cools at approximately 3°C per minute. To make a cup of tea, water should be a maximum of 85°C. How long should you wait after a kettle has boiled before making a cup of tea?

9 A metal object has mass 345 kg and volume 0.15 m³. Calculate the density of the object.

10 One day Ian runs 8 km. The next day he runs 10 km. Work out the percentage increase in the distance he runs.

11 In a restaurant, 13 of the 25 customers order the set menu. Work out the proportion of customers who do not order the set menu.

12 A DJ runs a radio programme between 9.00 am and 10.30 am. Here is the schedule.

9.00 am	News and Weather
9.15 am	80s Music
10.00 am	News
10.05 am	Weather
10.08 am	Quiztime
10.20 am	Showbiz Roundup

What proportion of the schedule is taken up with news and weather?

13 A snail crawls at $\frac{4}{5}$ of a metre per hour. How long will it take to travel the length of a garden measuring 20 m?

14 Josie and Charlie share 50 marbles in the ratio 1 : 4.

   a How many more marbles does Charlie have than Josie?

   b Write a function relating the number of marbles Josie has (J) to the number Charlie has (C).

▦ You **may** use a calculator for these questions.

15 What percentage of this circle is shaded? Give your answer to 2 significant figures.

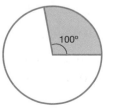

100°

119

16 In a shop the normal price of a sofa is £1200. In a pre-Christmas sale all prices are reduced by 20%.

After Christmas a further 10% is taken off sale prices.

What is the price of the sofa now?

17 A map uses a scale of 1:50 000. A particular road is 3 km long. Work out the length of the road on the map. Give your answer in cm.

18 Henry invests £1500 in a savings account. He is paid 2% per year in compound interest.

How much will Henry have in the account at the end of 3 years?

19 One in four people who are contacted by a charity donate to the charity. If the charity contacts 32 000 people in a year, how many will donate?

20 The train track from London to Edinburgh is 393 miles long. If a train travels at a speed of 125 mph, how long will the journey take? Give your answer in hours and minutes.

21 A café charges the following prices:

Coffee   £2.50

Tea   £1.90

Sandwiches   £5.30

All prices are increased by 5%.

Mr Angle always buys 1 coffee, 1 tea and 2 sandwiches. What is the new price of his order?

22 The number of people using the facilities in a sports centre at 10.00 am one morning is recorded in a table as follows.

Facility	Number of customers
Swimming pool	37
Gym	15
Squash courts	4
Café	19

What percentage of the customers are using the gym?

23 A savings account offers 4.5% simple interest.

If Allan puts £3000 into an account, how much will he have in the account at the end of 5 years?

24 There are 30 children in a class. The ratio of boys to girls is 2:1. Two boys leave and another three girls join the class. What is the ratio of boys to girls now?

25 A concert is attended by 3450 people.

The ratio of men to women is 7:6.

The ratio of women to children is 15:2.

How many men are at the concert?

26 A mobile phone company uses this graph to calculate the cost of mobile phone bills.

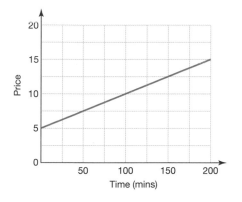

Erin says 'the cost of the bill is in direct proportion to the number of minutes of calls you make'.

Is she correct? Explain your answer.

27 The time to cook a turkey is given as 25 minutes per 450 g, plus an extra 25 minutes.

Is the relationship between the weight of the turkey and the time taken to cook it:

  i direct proportion

 ii inverse proportion

iii neither?

Explain your answer.

28 A crocodile can run at a speed of 17 km per hour over short distances. How long will it take to run 50 m? Give your answer to the nearest second.

29 The table shows the people using a hotel on one day.

	Adults	Children
Female	35	12
Male	60	37

Alex says 'the proportion of adults and children who are male is the same, since $37 - 12 = 25$ and $60 - 35 = 25$'. Is she correct? Explain your answer.

Geometry and measures

Measuring and drawing angles

An **acute angle** is less than 90°.

An **obtuse angle** is greater than 90° but less than 180°.

A **right angle** is exactly 90°.

A **reflex angle** is more than 180°.

You can use letters to name angles. This is angle *ABC*.

You use a **protractor** to measure and draw angles.

The angle is always at the middle letter. This is angle *ABC*. The notation ∠*ABC* can also be used to refer to angle *ABC*.

Measuring angles

To measure an angle:

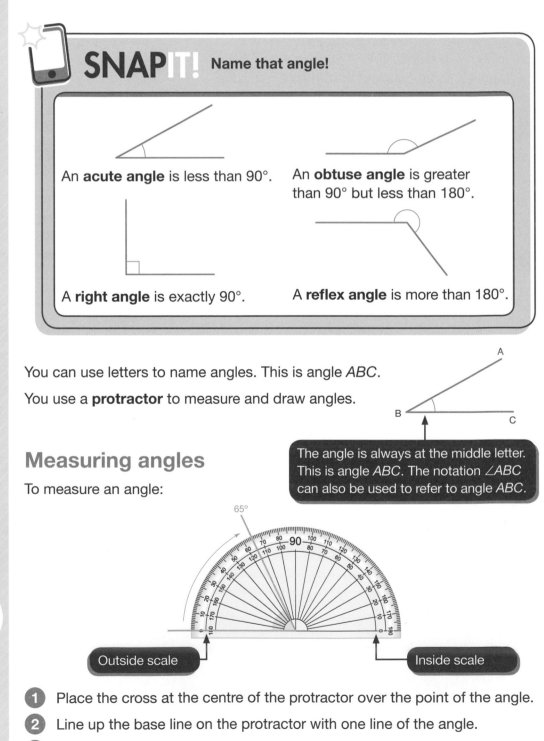

65°

Outside scale

Inside scale

1. Place the cross at the centre of the protractor over the point of the angle.

2. Line up the base line on the protractor with one line of the angle.

3. Use the outside scale to measure clockwise turns. Use the inside scale to measure anticlockwise turns.

WORKIT!

Measure reflex angle *DEF*.

Smaller angle = 115°

Angle DEF = 360° − 115° = 245°

Measure the smaller angle using the inside scale (start from 0°). Subtract the smaller angle from 360°. (See angle facts on page 124.)

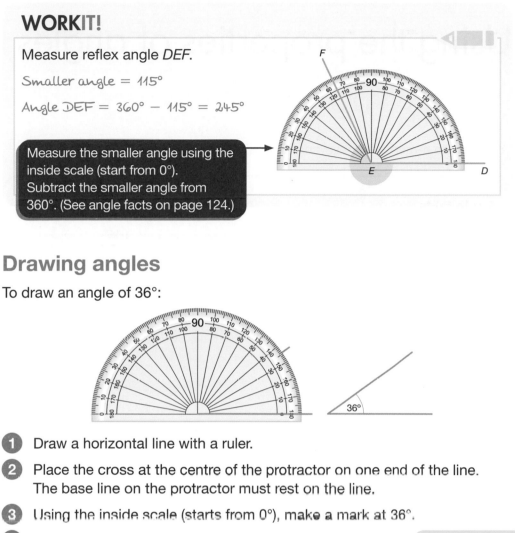

Drawing angles

To draw an angle of 36°:

1 Draw a horizontal line with a ruler.

2 Place the cross at the centre of the protractor on one end of the line. The base line on the protractor must rest on the line.

3 Using the inside scale (starts from 0°), make a mark at 36°.

4 Join the mark to the end of the line with a ruler.

5 Label the angle.

DOIT!

Draw and label angle *ABC* = 140°.
Mark on your diagram angle *ABC* = 220°.

CHECKIT!

1 a Measure the size of the angle marked *x*.

b What type of angle is this?

2 Draw and label these angles.

a angle *XYZ* = 118°

b angle *ABC* = 290°

c angle *LMN* = 42°

3 Here is triangle *PQR*. This diagram is accurately drawn.

a Mark with a letter *O* an obtuse angle.

b Measure the size of the angle *QRP*.

Using the properties of angles

Angle facts

You need to remember these angle facts.

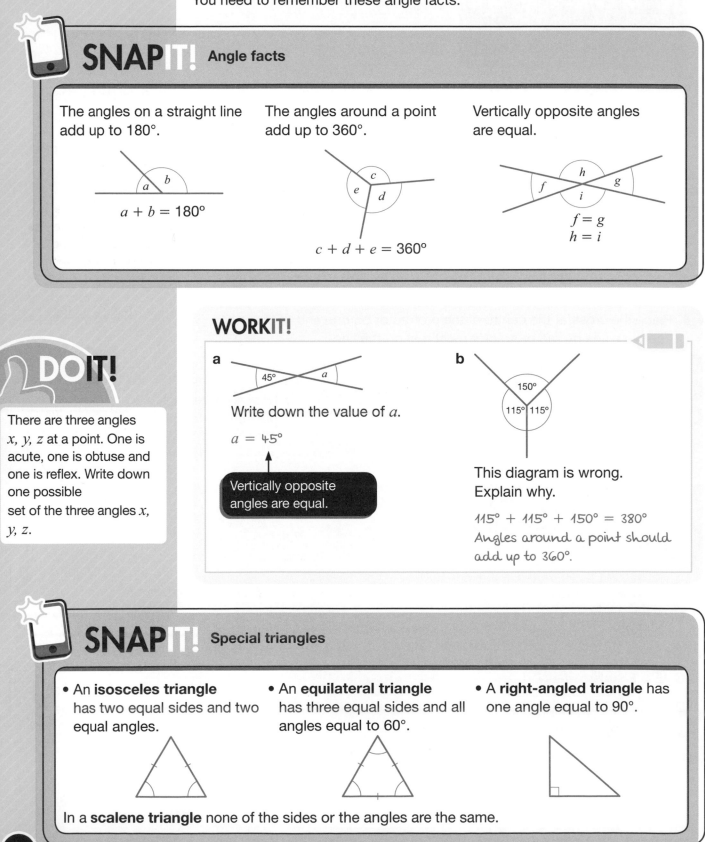

SNAPIT! Angle facts

The angles on a straight line add up to 180°.

$a + b = 180°$

The angles around a point add up to 360°.

$c + d + e = 360°$

Vertically opposite angles are equal.

$f = g$
$h = i$

DOIT!

There are three angles x, y, z at a point. One is acute, one is obtuse and one is reflex. Write down one possible set of the three angles x, y, z.

WORKIT!

a

45° a

Write down the value of a.

$a = 45°$

Vertically opposite angles are equal.

b

150°
115° 115°

This diagram is wrong. Explain why.

$115° + 115° + 150° = 380°$
Angles around a point should add up to 360°.

SNAPIT! Special triangles

- An **isosceles triangle** has two equal sides and two equal angles.

- An **equilateral triangle** has three equal sides and all angles equal to 60°.

- A **right-angled triangle** has one angle equal to 90°.

In a **scalene triangle** none of the sides or the angles are the same.

Angles in triangles

The **interior angles** of a triangle add up to 180°.

$$a + b + c = 180°$$

An **exterior angle** of a triangle is equal to the sum of the interior angles at the other two **vertices**.

A **vertex** (plural: **vertices**) is a point where two edges meet.

You can use letters to name lines and the sides of shapes. Here are line *BC* and triangle *DEF*.

DO IT!

One angle in an isosceles triangle is 70°. Show that there are two possible solutions for the other two angles.

One angle in an isosceles triangle is 72°. Work out the size of the smallest possible angle in the triangle.

WORKIT!

BA = *BC*

a Find the size of the angle marked *b*.

Angle *b* = 180° − 55° − 55° = 70°

b Give a reason for your answer.

Triangle *ABC* is isosceles.

Angle *BAC* = angle *BCA*.

The **base angles** of an isosceles triangle are the equal angles.

Write angle *BCA* on the diagram.

STRETCH IT!

The angles in triangle *PQR* are in the ratio 1 : 2 : 3

What type of triangle is *PQR*?

Angles in quadrilaterals

The interior angles of a quadrilateral add up to 360°.

$$a + b + c + d = 360°$$

SNAP IT! Opposite angles of a parallelogram

$$e = g$$
$$f = h$$

See page 134 for more on the properties of quadrilaterals.

WORKIT!

ABCD is a quadrilateral.

CDE is a straight line.

Work out the size of the angle marked *x*.

Give reasons for your answer.

Angles in a quadrilateral add up to 360°.

Angle *ADC* = 360° − 108° − 122° − 65° = 65°

Angles on a straight line add up to 180°.

Angle *x* = 180° − 65° = 115°

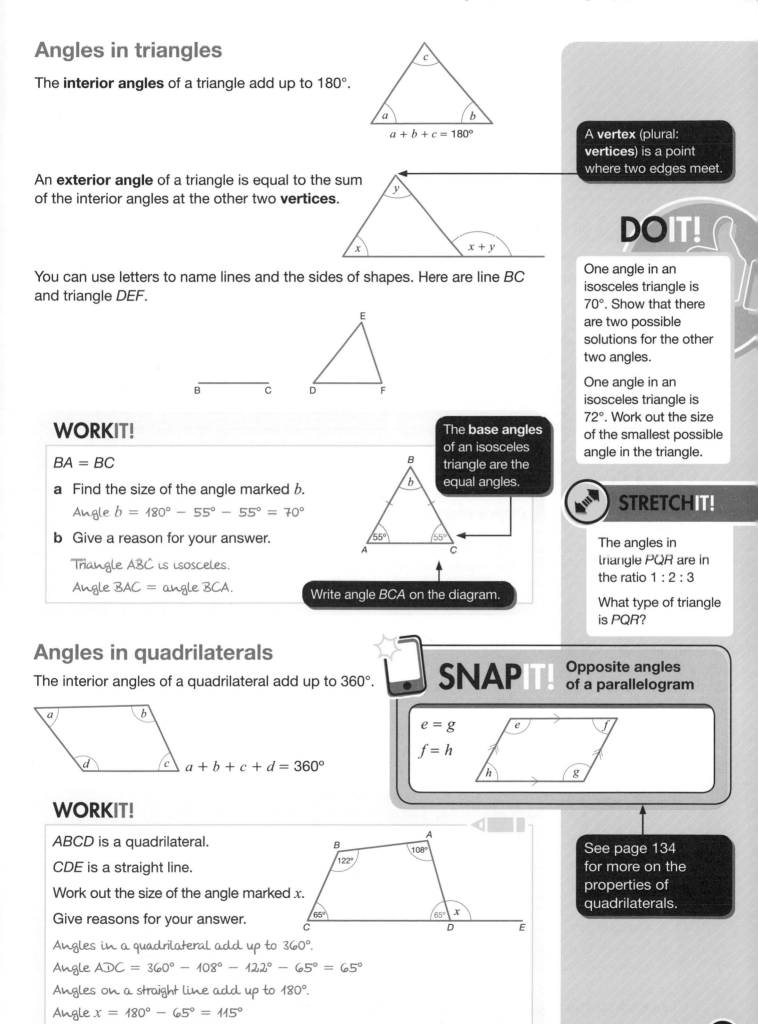

Parallel and perpendicular lines

Perpendicular lines meet at 90°. In this diagram, line *PQ* is perpendicular to line *RS*.

Parallel lines are lines that always remain the same distance apart. On diagrams they are marked with arrows.

You need to remember these angle facts about parallel lines.

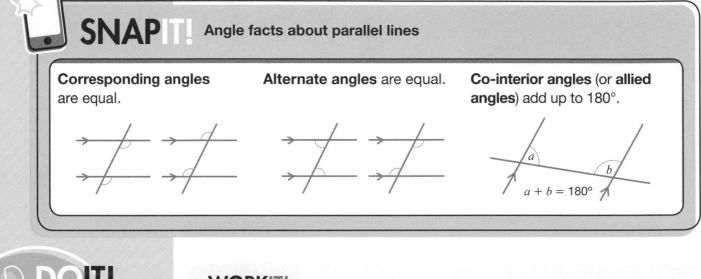

SNAP IT! Angle facts about parallel lines

Corresponding angles are equal.

Alternate angles are equal.

Co-interior angles (or **allied angles**) add up to 180°.

$a + b = 180°$

DO IT!

Here is parallelogram *ABCD*.

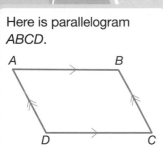

a Draw a line through *B* perpendicular to *DC*.

b Draw a line through the midpoint of *AB* parallel to *CB*.

WORKIT!

BC is a straight line.

Is the line *AB* parallel to the line *DC*?

Give a reason for your answer.

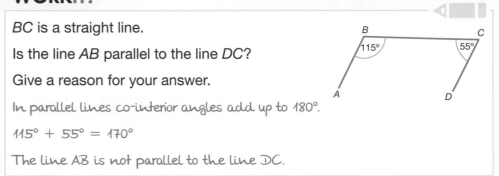

In parallel lines co-interior angles add up to 180°.

115° + 55° = 170°

The line AB is not parallel to the line DC.

Solving angle problems

When solving more involved angle problems you will need to use two or **more** angle properties. Make sure you give a **clear** reason for each step of your working.

DO IT!

Create a poster to show each of these angle properties you will use in angle problems.

a Angles on a straight line add up to 180°.

b Angles around a point add up to 360°.

c Vertically opposite angles are equal.

d Angles in a triangle add up to 180°.

e Base angles of an isosceles triangle are equal.

f Angles in a quadrilateral add up to 360°.

g Opposite angles of a parallelogram are equal.

h Corresponding angles are equal.

i Alternate angles are equal.

WORKIT!

PQR is a straight line.

PT = *QT*

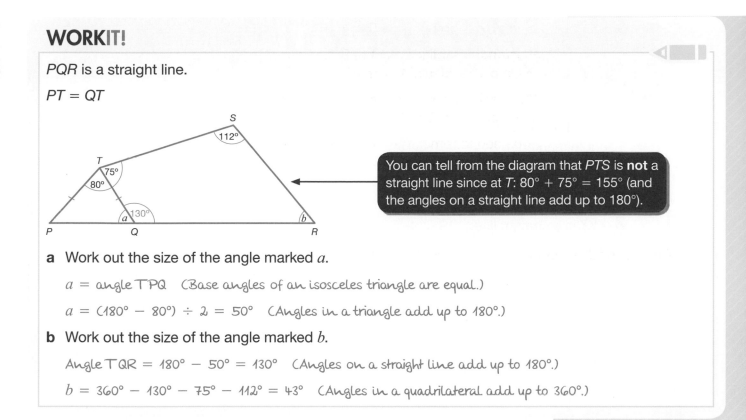

You can tell from the diagram that *PTS* is **not** a straight line since at *T*: 80° + 75° = 155° (and the angles on a straight line add up to 180°).

a Work out the size of the angle marked *a*.

a = angle TPQ   (Base angles of an isosceles triangle are equal.)

a = (180° − 80°) ÷ 2 = 50°   (Angles in a triangle add up to 180°.)

b Work out the size of the angle marked *b*.

Angle TQR = 180° − 50° = 130°   (Angles on a straight line add up to 180°.)

b = 360° − 130° − 75° − 112° = 43°   (Angles in a quadrilateral add up to 360°.)

You also need to be able to 'show that…'.

WORKIT!

ABCD is a parallelogram.

Show that triangle *ECB* is an isosceles triangle.

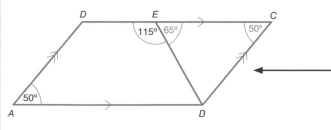

You are told that *ECB* is an isosceles triangle, so the base angles will be equal.
Work out all the angles of triangle *ECB* to show that two of the angles are equal.

Angle ECB = 50°   (Opposite angles of a parallelogram are equal.)

Angle CEB = 180° − 115° = 65°   (Angles on a straight line add up to 180°.)

Angle CBE = 180° − 50° − 65° = 65°   (Angles in a triangle add up to 180°.)

Base angles of an isosceles triangle are equal.

Angle CEB = angle CBE so triangle ECB is an isosceles triangle.

STRETCHIT!

The question could have been set as: '*ABCD* is a parallelogram. What type of triangle is triangle *ECB*?' Your working and conclusion would be exactly the same.

NAILIT!

For some angle problems, you may be given the angles in terms of letters and numbers, for example $x + 24°$. You will need to set up and solve equations to find the missing angles. If you're unsure how to do this, have a look at page 54.

CHECK**IT!**

1 Find the size of the angle marked a.

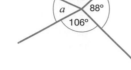

2 *XYZ* is a straight line.

WX = WY.

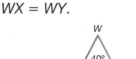

a **i** Work out the size of the angle marked a.

ii Give a reason for your answer.

b Find the size of the angle marked b.

3 Work out the value of x.

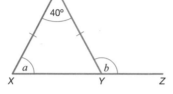

> Write an equation in terms of x using the fact that angles around a point add up to 360°.

4 *RST* is a straight line.

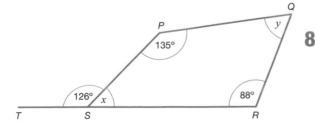

a **i** Find the size of the angle marked x.

ii Give a reason for your answer.

b Find the size of the angle marked y.

5 *AB* is parallel to *CD*. *EH* is a straight line.

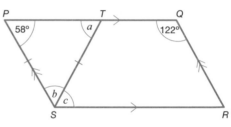

a Find the size of the angle marked x.

b **i** Find the size of the angle marked y.

ii Give a reason for your answer.

6 *PQRS* is a parallelogram.

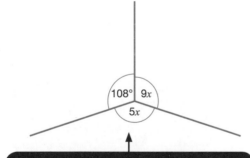

a Write down the angle marked a.

b Work out the size of the angle marked b.

c Work out the size of the angle marked c. Give a reason for your answer.

7 *AB* is parallel to *DC*. *DA* is parallel to *CB*. *AE = DE*. *EDC* is a straight line.

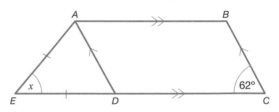

Work out the size of the angle marked x.

Give reasons for your answer.

8

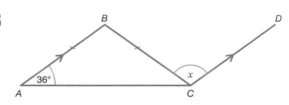

AB is parallel to *CD*. *AB = CB*.

Work out the size of the angle marked x. Give reasons for each stage of your working.

Using the properties of polygons

A **polygon** is a 2D shape with three or more straight sides.

In a **regular polygon** all the sides are equal **and** all the angles are equal; otherwise it is an **irregular polygon**.

You need to be able to recognise and name the polygons with 3 to 8 sides and the polygon with 10 sides.

A regular triangle would be an equilateral triangle (see page 124), and a regular quadrilateral would be a square (see page 134).

SNAP IT! **Name that polygon!**

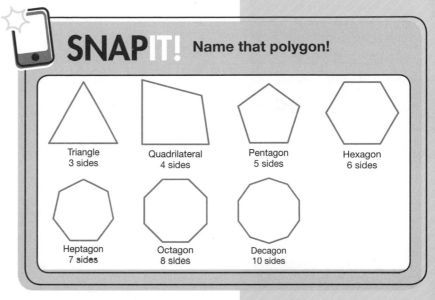

Triangle
3 sides

Quadrilateral
4 sides

Pentagon
5 sides

Hexagon
6 sides

Heptagon
7 sides

Octagon
8 sides

Decagon
10 sides

 STRETCH IT!

1 A hexagon can be divided into four triangles when all the diagonals are drawn from one vertex (point).

Complete these statements:

The angle sum of a triangle is _____°

Sum of interior angles of a hexagon = 4 × _____° = _____°

2 Think about dividing other polygons in the same way. Complete this table.

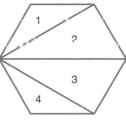

Polygon	Number of sides (n)	Number of triangles formed	Sum of interior angles
Triangle	3	1	180°
Quadrilateral	4	2	
Pentagon	5		
Hexagon			720°
Heptagon		5	
Octagon			1080°
Decagon			

3 How many triangles are formed for an n-sided polygon?

4 Write down the sum of the interior angles of an n-sided polygon.

The sum of the **interior angles** of a polygon with n sides is given by: $180° \times (n - 2)$

For the polygon shown, $n = 5$.

So, sum of interior angles $= 180° \times (5 - 2)$

$= 180° \times 3 = 540°$

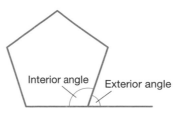

Interior angle   Exterior angle

You can use the fact that the angles on a straight line add up to 180° to work out an **exterior angle** if you know the interior angle, **or** to find the interior angle if you know the exterior angle.

At a vertex, interior angle + exterior angle = 180°

The sum of the exterior angles of any polygon is 360°.

x is an exterior angle. The exterior angles of a polygon add up to 360°. If a regular polygon has n sides, each exterior angle is 360° ÷ n.

y is an interior angle. Interior angle + exterior angle = 180°

WORKIT!

Here is a regular pentagon.

a Work out the size of the angle marked x.

$x = 360° \div 5 = 72°$

b Work out the size of the angle marked y.

$y = 180° - 72° = 108°$

If you know the exterior angle of a regular polygon, you can work out how many sides the polygon has.

WORKIT!

The interior angle of a regular polygon is 162°.

Work out how many sides the polygon has.

First work out the exterior angle.

Exterior angle = $180° - 162° = 18°$

Number of sides = $360° \div 18° = 20$

The polygon is regular so all the exterior angles are 18°, with sum 360°.

STRETCHIT!

Regular hexagons can fit together to form a pattern without any gaps or overlaps.

A square and an octagon can also fit together to form a pattern.

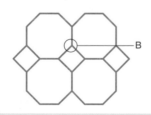

By considering the sum of angles at points A and B, give a reason why these polygons fit together as shown.

Can you fit together regular pentagons to form a pattern without any gaps or overlaps? Give a reason for your answer.

CHECKIT!

1 Work out the size of the exterior angle of a regular decagon.

2 The exterior angle of a regular polygon is 15°.

   a Work out how many sides the polygon has.

   b Work out the sum of the interior angles of the polygon.

3 The diagram shows a regular pentagon ABCDE.

The line EB is drawn.

Work out the size of angle CBE.

What can you say about triangle ABE?

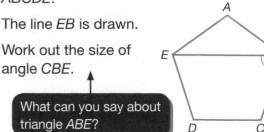

Using bearings

A **bearing** is an angle measured clockwise from North, used to describe a direction.

A bearing is always written using three figures.

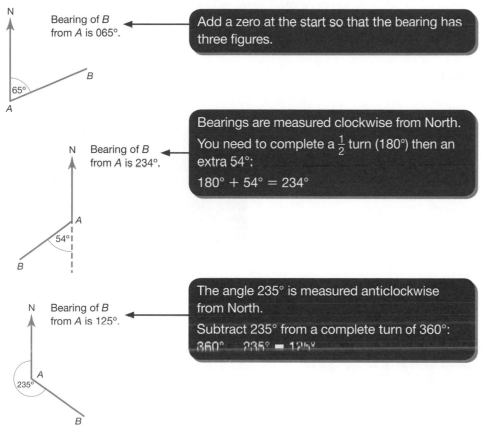

Bearing of *B* from *A* is 065°.

> Add a zero at the start so that the bearing has three figures.

Bearing of *B* from *A* is 234°.

> Bearings are measured clockwise from North. You need to complete a $\frac{1}{2}$ turn (180°) then an extra 54°:
> 180° + 54° = 234°

Bearing of *B* from *A* is 125°.

> The angle 235° is measured anticlockwise from North.
> Subtract 235° from a complete turn of 360°:
> 360° − 235° = 125°

You should know how to 'measure the bearing of…'. The term 'measure' is telling you to use a protractor. If you're unsure about using a protractor, have a look at page 122.

You can work out a reverse bearing by adding or subtracting 180°.

WORKIT!

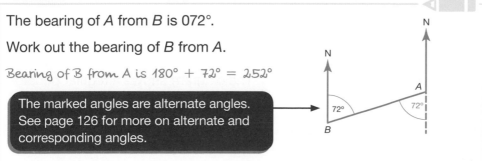

The bearing of *A* from *B* is 072°.

Work out the bearing of *B* from *A*.

Bearing of *B* from *A* is 180° + 72° = 252°

> The marked angles are alternate angles. See page 126 for more on alternate and corresponding angles.

NAILIT!

Bearing of *B* from *A*

'From *A*' means that you are 'at *A* facing North'.

You need to work out what angle you need to turn through to face *B*.

DOIT!

Compass points

Draw the eight points of a compass (N, NE, E, SE, S, SW, W and NW). Add the three-figure bearing of each point (for example, N 000°, E 090°).

To solve a bearings problem you may need to give a bearing between the points on a map or scale drawing (see page 103). You also need to be able to draw an accurate diagram.

WORKIT!

The diagram shows the positions of two buildings, P and Q, on a map.

A building R is on a bearing of 055° from building P.

Building R is also on a bearing of 280° from building Q.

Draw an accurate diagram to show the position of building R.

Label it R.

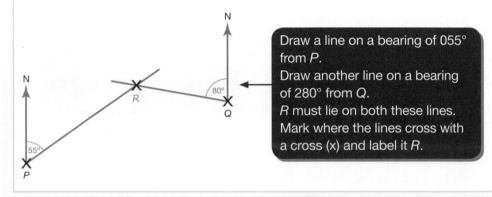

Draw a line on a bearing of 055° from P.
Draw another line on a bearing of 280° from Q.
R must lie on both these lines. Mark where the lines cross with a cross (x) and label it R.

✓ CHECKIT!

1

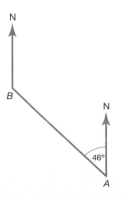

a Measure and write down the bearing of B from A.

b On the diagram, draw a line on a bearing of 115° from A.

2 The bearing of Q from P is 164°.

Work out the bearing of P from Q.

Draw a sketch of P and Q.

3 The diagram shows the positions of two boats, A and B.

Nathan says, 'The bearing of boat B from boat A is 046°'.

Kirsty says, 'The bearing of boat B from boat A is 314°'.

Who is correct? Give a reason for your answer.

Properties of 2D shapes

Lines of symmetry

A **line of symmetry** (or **mirror line**) is a line through a 2D shape such that each side is a mirror image of the other.

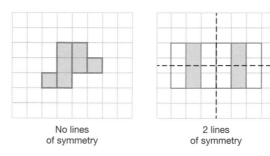

No lines
of symmetry

2 lines
of symmetry

You should know how to complete the drawing of a shape so that it has line symmetry.

You also need to be able to shade more squares to complete a drawing so that it has line symmetry.

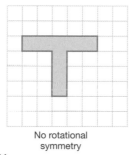

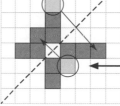

1 line of symmetry

> Each vertex (corner) of the shape has a mirror image the same distance away from the line of symmetry but on the other side of it.

Rotational symmetry

A shape has **rotational symmetry** if it can be turned and fitted onto itself within one full turn.

The **order of rotational symmetry** tells you how many times the shape fits onto itself.

If a shape fits onto itself **only once** in a full turn it does not have rotational symmetry.

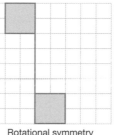

No rotational
symmetry

Rotational symmetry
of order 2

A shape that has rotational symmetry of order 2 looks the same after it has been rotated through a half turn.

> Each shaded square must have a mirror image if the pattern has the given line of symmetry. By comparing the two sides of the pattern, you can identify the squares that need to be shaded.

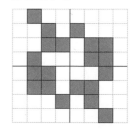

NAILIT!

You can use tracing paper to help you check for lines of symmetry and rotational symmetry.

1. Trace the shape.
2. Fold it in half along the line of symmetry. Check that both halves match up exactly.
3. Rotate the tracing paper to see how many times the shape fits onto itself in one full turn.

DOIT!

Rotating polygons

1. Find the order of rotational symmetry of a
   a. regular triangle (equilateral triangle)
   b. regular quadrilateral (square)
   c. regular pentagon
   d. regular hexagon.
2. What is the link between the order of rotational symmetry and the number of sides of a regular polygon?

Quadrilaterals

A **quadrilateral** is any four-sided shape.

You need to know (and remember!) the properties of these special quadrilaterals.

SNAP IT! Quadrilaterals

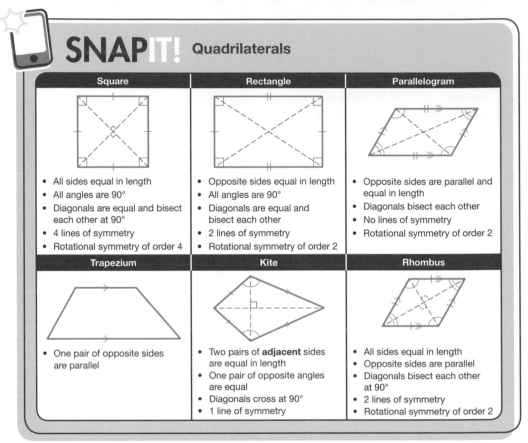

Square	Rectangle	Parallelogram
• All sides equal in length • All angles are 90° • Diagonals are equal and bisect each other at 90° • 4 lines of symmetry • Rotational symmetry of order 4	• Opposite sides equal in length • All angles are 90° • Diagonals are equal and bisect each other • 2 lines of symmetry • Rotational symmetry of order 2	• Opposite sides are parallel and equal in length • Diagonals bisect each other • No lines of symmetry • Rotational symmetry of order 2

Trapezium	Kite	Rhombus
• One pair of opposite sides are parallel	• Two pairs of **adjacent** sides are equal in length • One pair of opposite angles are equal • Diagonals cross at 90° • 1 line of symmetry	• All sides equal in length • Opposite sides are parallel • Diagonals bisect each other at 90° • 2 lines of symmetry • Rotational symmetry of order 2

Make sure that you can:
- write down the name of a quadrilateral given in diagram form
- draw any of the quadrilaterals on a squared grid
- name the quadrilaterals that have a specific property
- complete a quadrilateral, given information about the shape, on coordinate axes (see page 73).

CHECK IT!

1 a Complete this shape so that it has line symmetry.

  b Write down the name of the complete shape.

2 Here is a regular octagon.

  a Draw a line of symmetry on the octagon.

  b What is the order of rotational symmetry of the octagon?

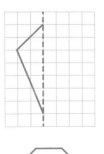

3 Complete the following sentences.

  a A rectangle has rotational symmetry of order _____.

  b A _____ has all sides equal and rotational symmetry of order 2.

  c A kite has _____ line of symmetry and _____ rotational symmetry.

  d The diagonals of a _____ and a _____ bisect each other at 90°.

Congruent shapes

Shapes are **congruent** if they have exactly the same shape and size.

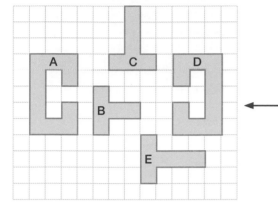

Shapes A and D are congruent.
These are reflected shapes
(see page 166).

Shapes C and E are congruent.
These are rotated shapes
(see page 167).

Shapes B, C and E are not congruent
as shape B is different.

If two triangles are congruent their angles are the same and corresponding sides are the same length.

The diagram shows a pair of congruent triangles X and Y.

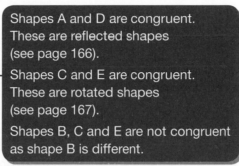

If you rotated triangle Y it would fit exactly onto triangle X.

Since triangles X and Y are congruent, you can write down that the length of side a is 8 cm and the size of angle b is 60°.

Two triangles are congruent when one of four conditions is true.

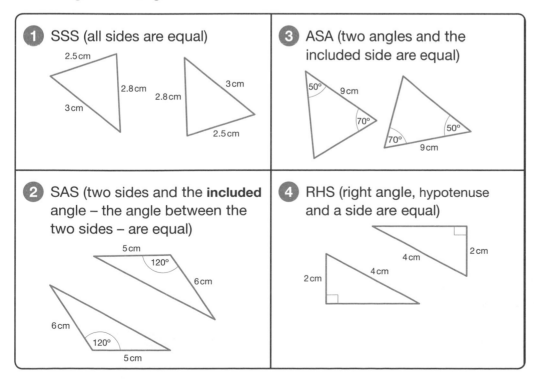

1. SSS (all sides are equal)
   2.5 cm, 2.8 cm, 3 cm, 2.8 cm, 3 cm, 2.5 cm

2. SAS (two sides and the **included** angle – the angle between the two sides – are equal)
   5 cm, 120°, 6 cm, 6 cm, 120°, 5 cm

3. ASA (two angles and the included side are equal)
   50°, 9 cm, 70°, 70°, 50°, 9 cm

4. RHS (right angle, hypotenuse and a side are equal)
   4 cm, 2 cm, 2 cm, 4 cm

NAIL IT!

The **hypotenuse** is the longest side and is opposite the right angle in a right-angled triangle.

WORKIT!

This pair of triangles is congruent. Explain why.

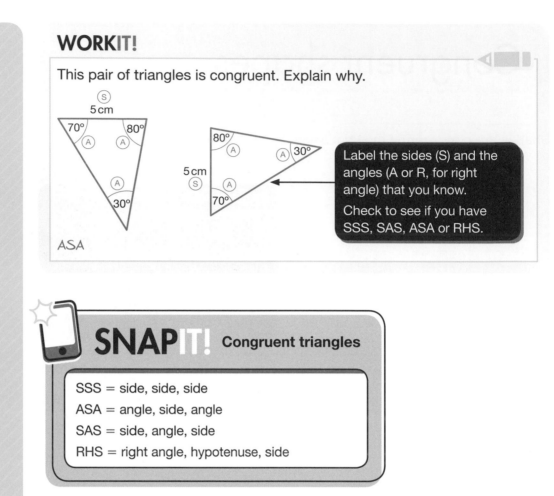

Label the sides (S) and the angles (A or R, for right angle) that you know.

Check to see if you have SSS, SAS, ASA or RHS.

ASA

SNAPIT! Congruent triangles

SSS = side, side, side

ASA = angle, side, angle

SAS = side, angle, side

RHS = right angle, hypotenuse, side

CHECKIT!

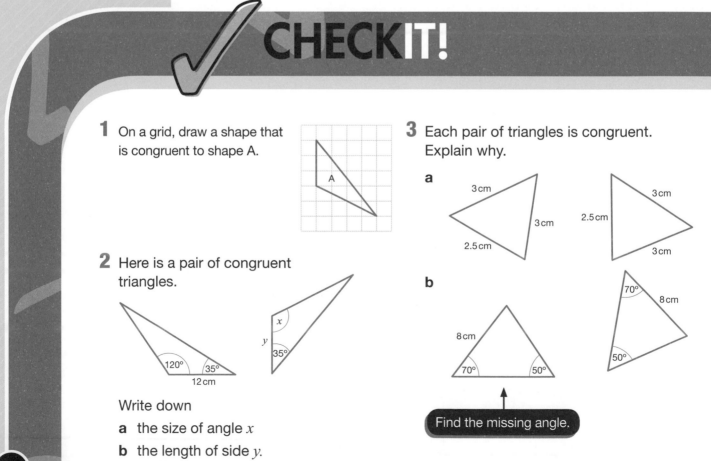

1 On a grid, draw a shape that is congruent to shape A.

2 Here is a pair of congruent triangles.

Write down

a the size of angle x

b the length of side y.

3 Each pair of triangles is congruent. Explain why.

a

b

Find the missing angle.

Constructions

To draw and measure a straight line accurately you need to use a ruler.

Here, line *XY* is 5.5 cm or 55 mm long.

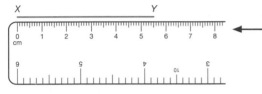

The ruler is marked off in cm with the smaller marks showing mm
(10 mm = 1 cm).
Line up the 0 on the ruler with the start of the line at *X*.
Measure to the nearest mm and include units in your answer.

Triangles and other 2D shapes

You can use a ruler and protractor to make an accurate drawing of triangle *ABC*.

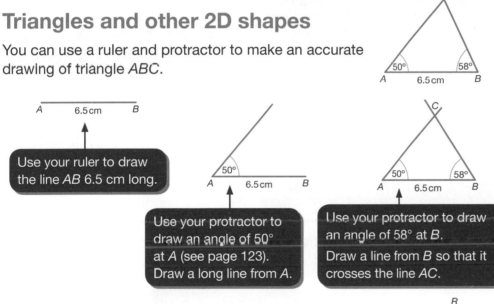

Use your ruler to draw the line *AB* 6.5 cm long.

Use your protractor to draw an angle of 50° at *A* (see page 123). Draw a long line from *A*.

Use your protractor to draw an angle of 58° at *B*. Draw a line from *B* so that it crosses the line *AC*.

If you are given the lengths of three sides, you should use a ruler and compasses to construct the triangle, for example *PQR* shown.

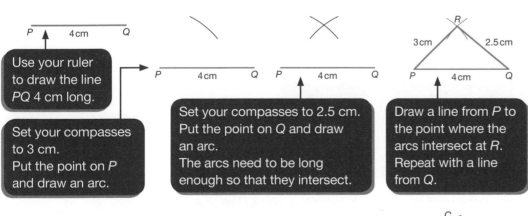

Use your ruler to draw the line *PQ* 4 cm long.

Set your compasses to 3 cm. Put the point on *P* and draw an arc.

Set your compasses to 2.5 cm. Put the point on *Q* and draw an arc. The arcs need to be long enough so that they intersect.

Draw a line from *P* to the point where the arcs intersect at *R*. Repeat with a line from *Q*.

You can now use a ruler, protractor and compasses to make accurate drawings of other 2D shapes including this quadrilateral.

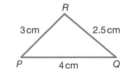
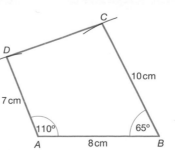

Perpendicular lines

A **perpendicular bisector** is a line that cuts another line exactly in half at right angles.

To construct the perpendicular bisector of the line AB you will need a ruler and compasses.

Open your compasses to more than half the length of line AB.
Put the point on A and draw an arc above and below the line.

Use your ruler to draw a line to join the points where the arcs intersect.
The line you have drawn is the perpendicular bisector.

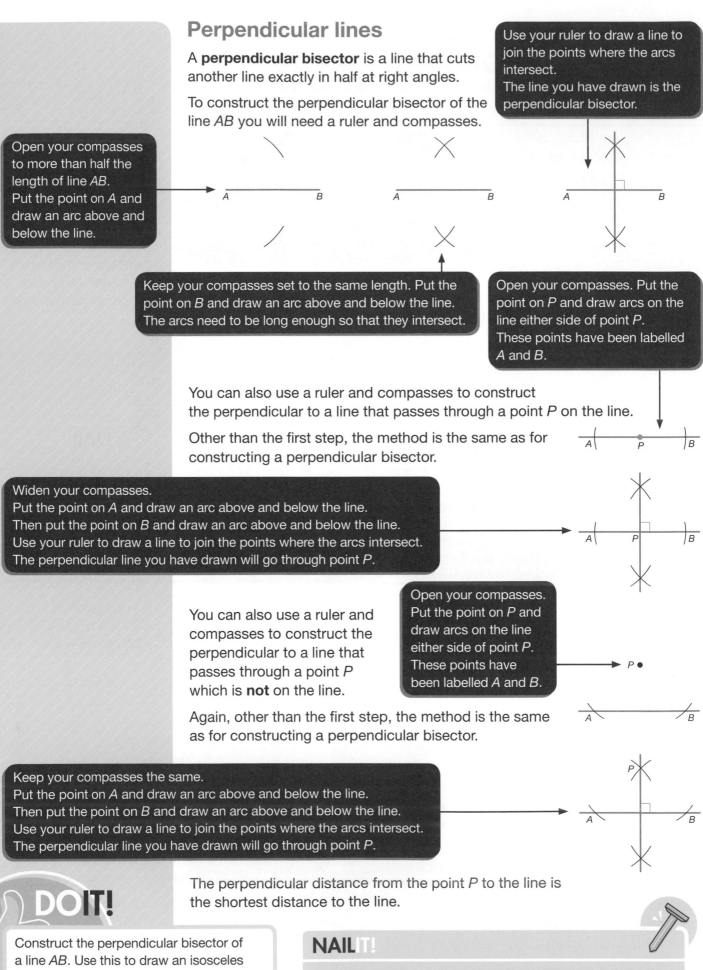

Keep your compasses set to the same length. Put the point on B and draw an arc above and below the line. The arcs need to be long enough so that they intersect.

Open your compasses. Put the point on P and draw arcs on the line either side of point P. These points have been labelled A and B.

You can also use a ruler and compasses to construct the perpendicular to a line that passes through a point P on the line.

Other than the first step, the method is the same as for constructing a perpendicular bisector.

Widen your compasses.
Put the point on A and draw an arc above and below the line.
Then put the point on B and draw an arc above and below the line.
Use your ruler to draw a line to join the points where the arcs intersect.
The perpendicular line you have drawn will go through point P.

You can also use a ruler and compasses to construct the perpendicular to a line that passes through a point P which is **not** on the line.

Open your compasses.
Put the point on P and draw arcs on the line either side of point P. These points have been labelled A and B.

Again, other than the first step, the method is the same as for constructing a perpendicular bisector.

Keep your compasses the same.
Put the point on A and draw an arc above and below the line.
Then put the point on B and draw an arc above and below the line.
Use your ruler to draw a line to join the points where the arcs intersect.
The perpendicular line you have drawn will go through point P.

The perpendicular distance from the point P to the line is the shortest distance to the line.

DO IT!

Construct the perpendicular bisector of a line AB. Use this to draw an isosceles triangle ABC.

NAIL IT!

If you are asked to find the **shortest distance** or construct the **shortest path** you should construct the perpendicular line.

Angles

An **angle bisector** is a line that cuts an angle exactly in half.

To construct the bisector of angle PQR you will need a ruler and compasses.

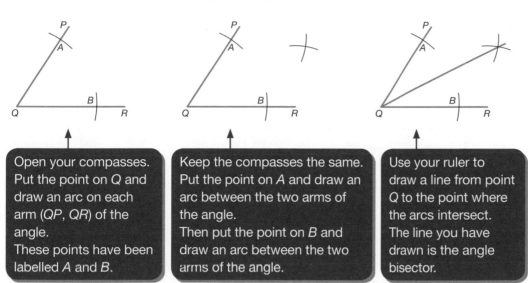

Open your compasses.
Put the point on Q and draw an arc on each arm (QP, QR) of the angle.
These points have been labelled A and B.

Keep the compasses the same.
Put the point on A and draw an arc between the two arms of the angle.
Then put the point on B and draw an arc between the two arms of the angle.

Use your ruler to draw a line from point Q to the point where the arcs intersect.
The line you have drawn is the angle bisector.

WORKIT!

Use ruler and compasses to construct an angle of 45° at A.

Construct the perpendicular bisector of the horizontal line.
Set the compasses to the distance AB.
Put the point on B and draw an arc on the perpendicular.
Use a ruler to draw a line from point A to the point where the arc cuts the perpendicular.

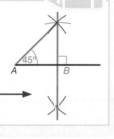

✓ CHECKIT!

In all questions, you must show all your construction lines.

1 Use ruler and compasses to construct an accurate drawing of triangle ABC.

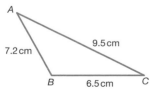

2 Rhombus ABCD has sides of length 6 cm. The diagonal AC = 5 cm.

Make an accurate drawing of rhombus ABCD.

3 Accurately draw the diagram below and use a ruler and compasses to construct the perpendicular to the line segment AB that passes through the point P.

4 Use ruler and compasses to construct the bisector of angle ABC.

5 Use ruler and compasses to construct an angle of 90°.

Drawing circles and parts of circles

DO IT!

Follow these steps to draw an arc of radius 4 cm and angle 50°.

1 Set your compasses to 4 cm.
Mark the centre with a cross (×) and draw part of the circle.

2 Place the centre of your protractor on the ×.
Make a mark at 50°.

3 Line up your ruler with the × and the mark.
Draw a line from the × to the arc.

DO IT!

After completing Question 1 of Check it! write sentences to describe the remaining parts of a circle.

You need to know the parts of a circle.

SNAP IT! Parts of a circle

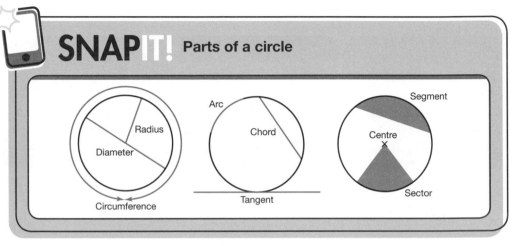

You should know how to draw a circle with a given radius or diameter.

For a circle of radius 4 cm work as shown.

Mark the centre with a cross (×).
Set your compasses to 4 cm.
Put the point of your compasses on the × and draw the circle.

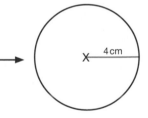

To draw the arc of a circle with a given radius and angle, you will need compasses and a protractor (see page 123).

CHECK IT!

1 Complete the following sentences.

a A _____ is a straight line that does not pass through the centre of a circle but does touch the circumference at each end.

b A _____ is a straight line that touches the outside of a circle at one point only.

c A _____ is a straight line through the centre of a circle that touches the circumference at each end.

d An _____ is part of the circumference of a circle.

e A _____ is a straight line from the centre to the edge of a circle that is half the length of the diameter.

f The part of a circle that has a chord and an arc as its boundary is called a _____.

2 a Draw a circle of diameter 9 cm.

b Draw a semicircle of radius 5 cm.

3 Draw an arc of radius 5 cm and angle 30°.

First work out the radius.

Loci

A **locus** (plural: **loci**) is a set of points that obey a certain rule.

You need to be able to construct loci using a ruler and compasses.

The locus of points that are **equidistant** (same distance) from a point *C* is a circle, centre *C*.

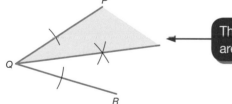

The locus of points that are 3 cm from point *C* is a circle, centre *C*, radius 3 cm.

The shaded region shows the locus of points that are **less than** 3 cm from point *C*.

The region of points that are **greater than** 3 cm from point *C* lies outside the circle.

The locus of points that are the same distance from two points, *A* and *B*, is the perpendicular bisector of the line *AB*.

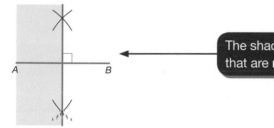

The shaded region shows the locus of points that are **nearer** to point *A* than point *B*.

DOIT!

Lines *AB* and *CD* are parallel and 4 cm apart.

1 Draw lines *AB* and *CD*.

2 Draw the locus of points that are the same distance from line *AB* and line *CD*.

3 Shade the region of points that are nearer to *CD* than *AB*.

The locus of points that are the same distance from two lines, *PQ* and *QR*, is the bisector of the angle *PQR*.

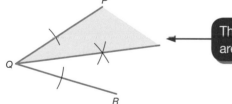

The shaded region shows the locus of points that are **nearer** to the line *PQ* than the line *QR*.

The locus of points that are 2 cm from the line *AB* consists of two lines parallel to *AB* and two semicircles (radius 2 cm).

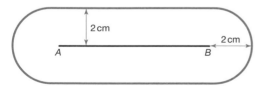

DOIT!

Using a scale of 1 cm : 20 cm draw a rectangular flowerbed 120 cm by 160 cm.

A fence is built around the flowerbed which is exactly 60 cm from each side.

Draw the fence.

To solve loci problems you may need to use scale drawings (see page 103) and/or bearings (see page 131).

WORKIT!

The diagram shows a field.

A goat is tethered to a post at X by a rope of length 6 m.

Shade the region that can be grazed by the goat.

Scale: 1 cm = 3 m

> The region is part of a circle of radius 6 m from X.
> 1 cm represents 3 m, so 2 cm represents 6 m.
> Set your compasses to 2 cm.
> Put the point on X and draw an arc between the two sides of the field.
> Shade the region.

You also need to be able to shade a region that satisfies more than one condition.

In rectangle ABCD, the shaded region represents the points that are more than 3 cm from point C **and** more than 4 cm from line AD.

WORKIT!

The diagram shows three radio masts P, Q and R.

Signals from P and Q have a range of 60 km.

Signals from R have a range of 80 km.

Shade the region in which the signals from all three masts can be received.

This diagram is not drawn to scale, but yours should be drawn with a scale of 1 cm = 20 km

> In your drawing, 1 cm represents 20 km
> P and Q: 3 cm represents 60 km
> Set your compasses to 3 cm.
> Put the point on P and draw a large arc. Put the point on Q and draw an arc.
> R: 4 cm represents 80 km
> Set your compasses to 4 cm.
> Put the point on R and draw an arc. Shade the region formed by the arcs.

CHECKIT!

1 Draw the locus of points that are exactly 4 cm from the line AB.

2 RST is a triangle.

Make an accurate copy of RST.

Shade the region inside the triangle that is less than 4 cm from point S **and** nearer to the line RS than to the line TS.

R
8.5 cm   10 cm
S   7.5 cm   T

3 Lighthouse L is on a bearing of 070° from lighthouse M.

Lighthouse L is 40 km from M.

a Construct a scale drawing to show the positions of lighthouses L and M. Use a scale of 1 cm = 10 km.

A boat sails so that it is always the same distance from lighthouses L and M.

b Draw the route taken by the boat.

Perimeter

The perimeter of a shape is the distance round the outside.

To calculate perimeter, add together the lengths of all the sides.

Sometimes you will need to calculate missing lengths before calculating perimeter.

WORKIT!

Calculate the perimeter of this shape made up of rectangles.

Perimeter = 12 + 9 + 12 + 9
= 42 cm

> The red lines must have a total length of 12 cm since the height of the whole shape is 12 cm. The blue lines must have a total length of 9 cm since the width of the whole shape is 9 cm.

12 cm

9 cm

You can use the properties of shapes to help you find their perimeter.

Shape	Property		Formula
Equilateral triangle	All sides are the same length.	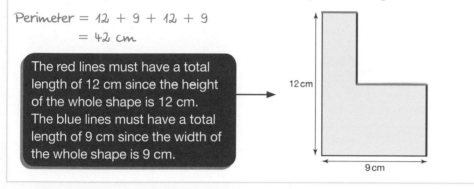	Perimeter = $3a$
Isosceles triangle	Two of the sides are equal in length.		Perimeter = $2a + b$
Square	All four sides have equal length.		Perimeter = $4a$
Rhombus	All four sides have equal length.		Perimeter = $4a$
Rectangle	Two pairs of sides are equal.		Perimeter = $2a + 2b$

Shape	Property		Formula
Parallelogram	Two pairs of sides are equal.	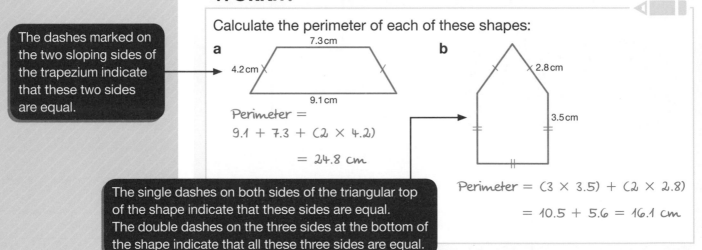	Perimeter = $2a + 2b$
Kite	Two pairs of sides are equal.		Perimeter = $2a + 2b$

WORKIT!

Calculate the perimeter of each of these shapes:

a

7.3 cm
4.2 cm
9.1 cm

Perimeter =
9.1 + 7.3 + (2 × 4.2)

= 24.8 cm

b

2.8 cm
3.5 cm

Perimeter = (3 × 3.5) + (2 × 2.8)

= 10.5 + 5.6 = 16.1 cm

> The dashes marked on the two sloping sides of the trapezium indicate that these two sides are equal.

> The single dashes on both sides of the triangular top of the shape indicate that these sides are equal.
> The double dashes on the three sides at the bottom of the shape indicate that all these three sides are equal.

Circumference

The perimeter of a circle is called the **circumference**.

STRETCHIT!

The Greek letter Pi (π) represents the area of a circle with a radius of 1 unit
π = 3.14159265…

NAILIT!

Circle vocabulary

Circumference = distance round the edge

Diameter = distance across the circle through the centre

Radius = distance from the centre to the circumference

Arc = part of the circumference

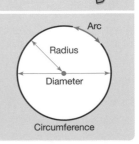

Arc
Radius
Diameter
Circumference

SNAPIT! Circumference

Circumference = $2\pi r = \pi d$

where r = radius and d = diameter

NAILIT!

Sometimes you need to give the answer rounded to a number of decimal places or significant figures – if this is the case, use your calculator to work out the answer. To give the exact answer, leave π in your answer.

WORKIT!

Calculate the circumference of this circle to 2 significant figures.

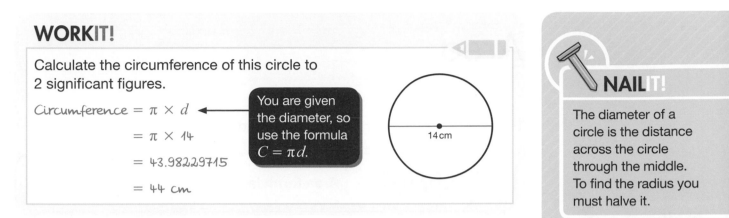

$Circumference = \pi \times d$

You are given the diameter, so use the formula $C = \pi d$.

$= \pi \times 14$

$= 43.98229715$

$= 44 \ cm$

The perimeter of a semicircle or a quarter-circle can be calculated by using parts of the circumference.

Semicircles

Length of curved edge $= \frac{1}{2} \times 2\pi r = \pi r$

The curved edge has half the length of the circumference of the circle.

Perimeter $= \pi r + 2r$

The straight edge is the diameter, which is double the radius.

Quarter-circles

Length of curved edge $= \frac{1}{4} \times 2\pi r = \frac{1}{2}\pi r$

The curved edge has one-quarter of the length of the circumference of the circle.

Perimeter $= \frac{1}{2}\pi r + 2r$

Each of the straight edges is a radius.

CHECKIT!

1 Work out the perimeter of a square with sides of length 7.2 cm.

2 Calculate the perimeter of this shape. You may assume all angles are right angles.

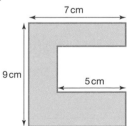

3 The perimeter of a semicircle can be calculated using the formula: $k\pi + b$. If the semicircle has radius 4 cm what are the values of k and b?

4 An athletics track has two semicircular ends with radius 15 m. They are connected by straight tracks of 100 m. Find the exact distance around the racetrack.

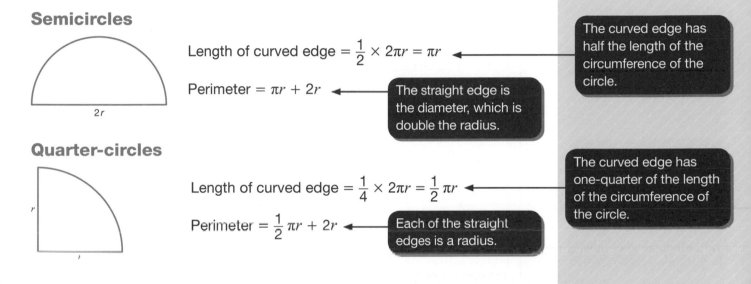

5 A cake is made in the shape of a quarter circle with radius 32 cm. Ribbon is tied around the perimeter of the cake. The ribbon has an overlap of 5 cm.

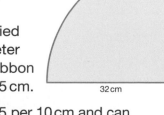

Ribbon costs £0.15 per 10 cm and can only be bought to the nearest 10 cm. Work out the cost of the ribbon for the cake.

Area

The **area** of a shape is the space it takes up in two dimensions.

Area is two-dimensional, so is measured in 'lengths squared' – for example, cm^2, m^2, km^2, $inches^2$.

SNAP**IT!** Area formula

Area of rectangle
= base × height
= $b \times h$

Area of parallelogram
= base × height
= $b \times h$

Area of triangle
= $\frac{1}{2}$ × base × height
= $\frac{1}{2} \times b \times h$

Area of trapezium
= $\frac{1}{2}$ (side a + side b) × height
= $\frac{1}{2} (a + b) \times h$

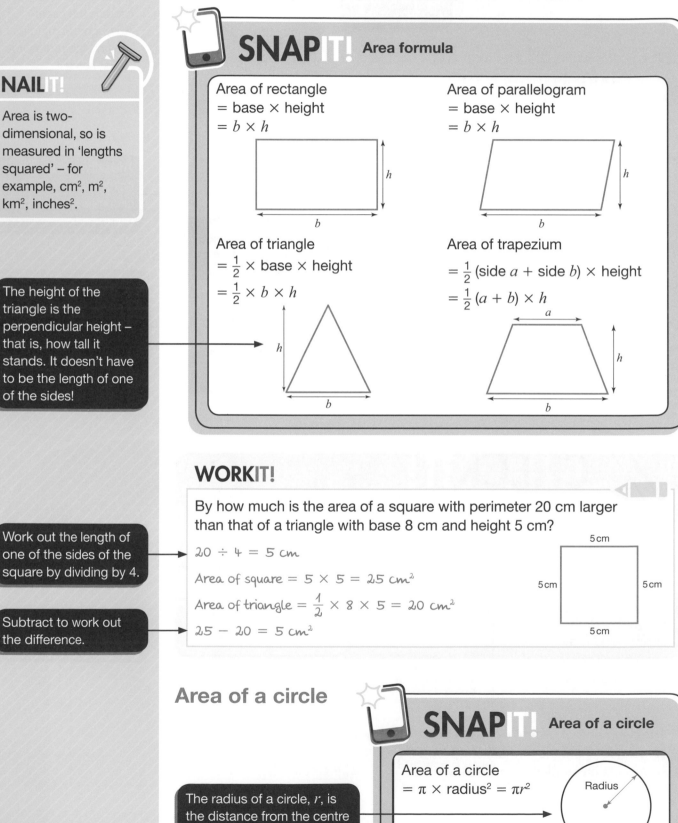

The height of the triangle is the perpendicular height – that is, how tall it stands. It doesn't have to be the length of one of the sides!

WORK**IT!**

By how much is the area of a square with perimeter 20 cm larger than that of a triangle with base 8 cm and height 5 cm?

$20 \div 4 = 5$ cm

Area of square = $5 \times 5 = 25$ cm^2

Area of triangle = $\frac{1}{2} \times 8 \times 5 = 20$ cm^2

$25 - 20 = 5$ cm^2

5 cm
5 cm   5 cm
5 cm

Work out the length of one of the sides of the square by dividing by 4.

Subtract to work out the difference.

Area of a circle

SNAP**IT!** Area of a circle

Area of a circle
= π × radius² = πr^2

Radius

The radius of a circle, r, is the distance from the centre of a circle to the perimeter.

WORKIT!

Work out the exact area of a circle with diameter 10 cm.

$10 \div 2 = 5 \, cm$ ← First work out the length of the radius.

$\pi \times 5^2 = \pi \times 25$

$\qquad = 25\pi \, cm^2$ ← Include π if you need an exact solution.

STRETCHIT!

Write down formulae for calculating the area of:

a a semicircle

b a quarter-circle.

DOIT!

Make some revision cards with pictures of shapes and the formula for finding the area of each shape. Learn them and ask someone to test you.

Composite shapes

A **composite** shape is one that is made up of more than one simple shape.

To find the area of a composite shape, cut it up into shapes with areas you know how to calculate.

For example:

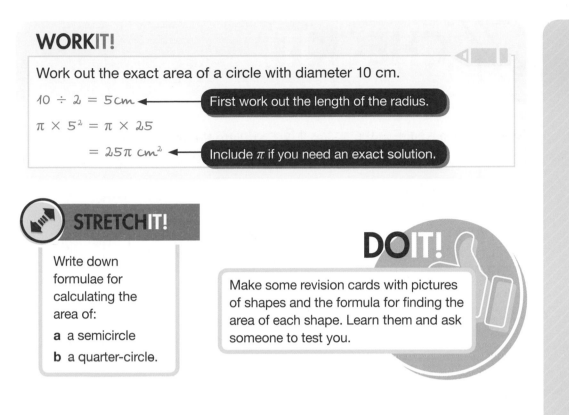

Area = $\frac{1}{2} \times b \times h$

can be split into

Area = $b \times h$

Make sure all your measurements are in the same units and find any unknown lengths before you start your calculation.

Sometimes it is easier to calculate a 'missing' area and subtract it from a larger area.

For example, to work out the area of this shape:

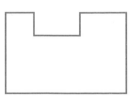

you could work out
Area A – Area B.

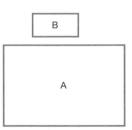

B

A

CHECK**IT!**

1 Work out the area of each of these shapes.

a

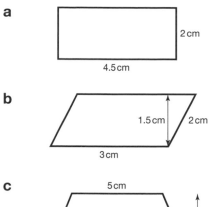

2 cm
4.5 cm

b

1.5 cm 2 cm
3 cm

c

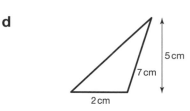

5 cm
6 cm
4 cm
9 cm

d

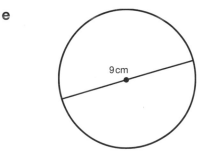

5 cm
7 cm
2 cm

e

9 cm

Give your answers to 1 decimal place.

2 Work out the area of a square with perimeter 12 cm.

3 Calculate the fraction of this trapezium that is shaded.

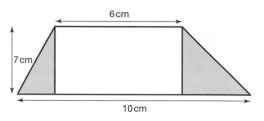

6 cm
7 cm
10 cm

4 The rectangles in this shape have equal areas. Work out the total area of the shaded rectangles.

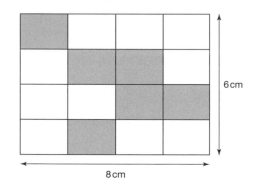

6 cm
8 cm

5 The diagram shows a square enclosing four identical circles.

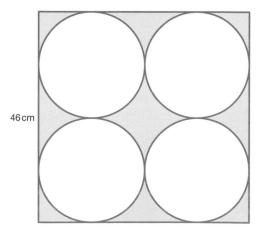

46 cm

Work out the area of the shaded region. Give your answer to 1 decimal place.

Sectors

A **sector** is a 'slice' of a circle taken from the centre of a circle.

Each slice of pizza is a 'sector'.

Area of a sector

To calculate the area of a sector, work out what fraction of the circle you are calculating and multiply the area of the circle by this fraction.

WORKIT!

Calculate the exact area of the sector.

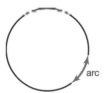
12 cm
60°

$$Area = \frac{60}{360} \times \pi \times 12^2$$
$$= \frac{1}{6} \times \pi \times 144$$
$$= 24\pi$$

Cancel the fraction down if you can.

SNAPIT!

Sector and arc

Area of a sector
$$= \frac{\theta}{360} \times \pi r^2$$

θ

where θ is the angle of the sector.

Length of arc $= \frac{\theta}{360} \times 2\pi r$

Arc length

The **arc** of a circle is part of the circumference.

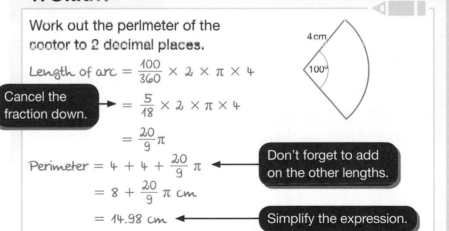
arc

To calculate the length of an arc, work out what fraction of the circle you are calculating and multiply the circumference by this fraction.

WORKIT!

Work out the perimeter of the sector to 2 decimal places.

4 cm
100°

$$Length\ of\ arc = \frac{100}{360} \times 2 \times \pi \times 4$$

Cancel the fraction down.

$$= \frac{5}{18} \times 2 \times \pi \times 4$$
$$= \frac{20}{9}\pi$$

$$Perimeter = 4 + 4 + \frac{20}{9}\pi$$

Don't forget to add on the other lengths.

$$= 8 + \frac{20}{9}\pi\ cm$$
$$= 14.98\ cm$$

Simplify the expression.

CHECKIT!

1 Work out the area and perimeter of a semicircle with diameter 10 cm.

Give your answers to 1 decimal place.

2 Calculate the exact area of this sector.

4 cm

3 A semicircular flower bed has a radius of 3 m. Angela buys topsoil to spread over the bed.

Bags of topsoil cost £14.99 and cover 2 m² each. How much will the topsoil cost? You must show all your calculations.

3D shapes

You need to be able to recognise and name these 3D shapes.

SNAPIT! 3D shapes

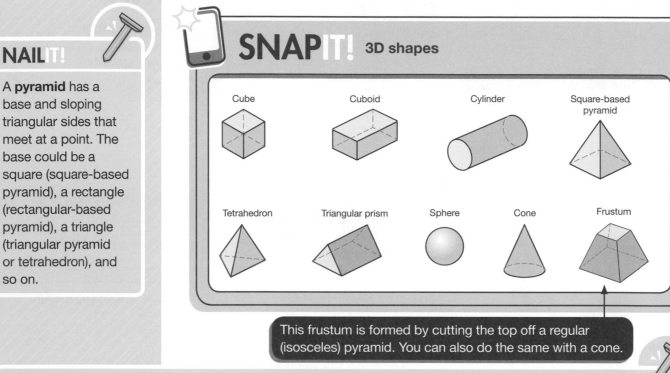

Cube · Cuboid · Cylinder · Square-based pyramid · Tetrahedron · Triangular prism · Sphere · Cone · Frustum

This frustum is formed by cutting the top off a regular (isosceles) pyramid. You can also do the same with a cone.

NAILIT!

A **prism** has the same **cross-section** through its length. This means it has two identical ends. A triangular prism has two (congruent) triangular ends and its cross-section is a triangle. The illustration shows a hexagonal prism – its cross-section is a hexagon.

A **frustum** is a slice of a cone or pyramid (any type), contained by two parallel planes. This means the top face of the frustum is parallel with the bottom face.

STRETCHIT!

Complete the following table.

3D shape	Faces	Edges	Vertices
Cube	6	12	
Cuboid			
Square-based pyramid			
Tetrahedron			
Triangular prism			
Hexagonal prism			

Properties of 3D shapes

Faces, edges and vertices

3D shapes have **faces**, **edges** and **vertices**.

A face is a flat surface of a 3D shape.

An edge is the line where two faces meet.

A **vertex** (plural: vertices) is the point where edges meet.

This tetrahedron has 4 faces, 6 edges and 4 vertices.

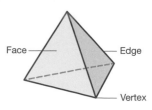

Face — Edge — Vertex

Planes of symmetry

A **plane of symmetry** divides a 3D shape into two equal halves which are mirror images of each other.

This isosceles triangular prism has 2 planes of symmetry.

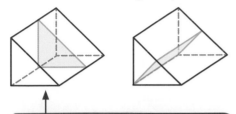

> Imagine cutting through the shape along the plane of symmetry. Would you end up with two equal halves?

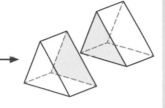

STRETCH IT!

This cuboid has no square faces. It has three planes of symmetry. Sketch the cuboid and mark on all the planes of symmetry. (You can use separate diagrams for each one.)

This cuboid has two opposite faces that are square. Sketch the cuboid and mark on all the planes of symmetry.

Drawing 3D shapes

You can draw a 3D shape using isometric paper.

You need to make sure that the isometric paper is the right way up! (Vertical lines down the page and no horizontal lines.)

This diagram represents a shape made using four cm cubes.

Nets

A **net** of a 3D shape is a 2D pattern that can be folded up to make the shape.

These are two possible nets for a cube.

You should be able to draw an accurate full-size net of a simple 3D shape.

STRETCH IT!

There are 11 possible nets for a cube.

Can you sketch all 11?

NAIL IT!

Make sure that you can also make an accurate full-size drawing of 3D shapes on isometric paper.

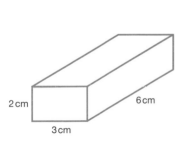

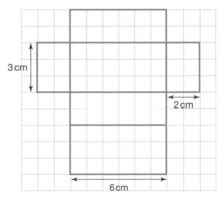

For some 3D shapes you will need to use a ruler and compasses.

Plans and elevations

Plans and **elevations** are 2D drawings of 3D shapes when viewed from different directions.

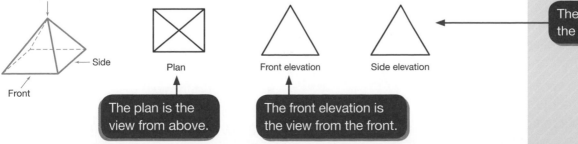

> The side elevation is the view from the side.

> The plan is the view from above.

> The front elevation is the view from the front.

You need to be able to draw the plan and elevations of a 3D shape.

WORKIT!

The diagram shows a solid made from 6 identical cubes.
On the grid, draw the plan, front and side elevations of the solid.

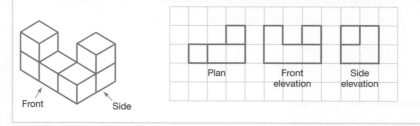

Front   Side

Plan   Front elevation   Side elevation

Given the plan and elevations of a 3D shape, you can sketch the shape.

WORKIT!

Here are the plan, front elevation and side elevation of a 3D shape.
Draw a sketch of the 3D shape.

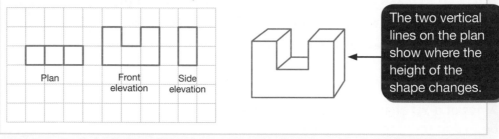

Plan   Front elevation   Side elevation

> The two vertical lines on the plan show where the height of the shape changes.

Given the plan and front elevation of a 3D shape, you also need to be able to draw the side elevation and then sketch the 3D shape.

DO IT!

Use cubes to build the 3D shape in the Work it! View your shape from above, from the front, and from the side.

CHECKIT!

1 Here is a cuboid.

a Shade in a rectangular face.

b Shade in the face *BCHG*.

c Kelli says 'A cuboid has nine edges'. Kelli is wrong. Explain what mistake she has made.

2 The diagram shows a square-based pyramid.

Draw an accurate net of the pyramid. Use a grid of cm squares.

6 cm

4 cm

4 cm

3 Each of these nets will form a 3D shape. Draw a sketch of each 3D shape.

a   b

4 The diagram shows a solid made from four identical cubes.

On a grid, draw the plan, front and side elevations.

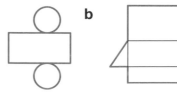

Front

5 Here are the plan and front elevation of a 3D shape.

Plan   Front elevation

a On a grid, draw the side elevation of the 3D shape.

b Draw a sketch of the 3D shape.

Volume

The volume of a shape is the space it takes up in three dimensions.

NAILIT!

Volume is three-dimensional, so is measured in 'lengths cubed' – for example, cm^3, m^3, km^3, $inches^3$.

SNAP IT! Volumes of shapes

Volume of cuboid = length × width × height

length
height
width

Volume of sphere = $\frac{4}{3}$ π radius^3 = $\frac{4}{3}\pi r^3$

r

Volume of cylinder = π × radius^2 × height
= $\pi r^2 h$

radius
height

Volume of a prism
= area of cross-section × length

area of cross-section
length

A **prism** is a three-dimensional shape with uniform cross-section. This means you could cut a slice and the cross-section would still be the same.
Note that a **cylinder** is a prism – it has a circular cross-section.

SNAP IT! Volume of pyramid and cone

Volume of a pyramid
= $\frac{1}{3}$ × area of base × height

height
area of base

Volume of cone
= $\frac{1}{3}$ × π × radius^2 × height
= $\frac{1}{3}\pi r^2 h$

height
radius

A **pyramid** is any three-dimensional shape that goes up to a point. This is a square-based pyramid, but the formula works for a pyramid with any shape base.

STRETCHIT!

If you need to find the volume of a regular frustum, remember that its top and bottom faces are parallel. So it is a regular (isosceles) pyramid with the top cut off.

Say the height of the original pyramid or cone was p, and the height of your frustum is f. You need to:

1 Work out the volume of the pyramid with height p. (Its base is the bottom of the frustum.)

2 Subtract the volume of the pyramid with height $p - f$. (Its base is the top of the frustum.)

Volume of composite shapes

To find the volume of a composite shape, split the shape into two or more 3D shapes with volumes that you can find.

WORKIT!

Workout the volume of this shape.

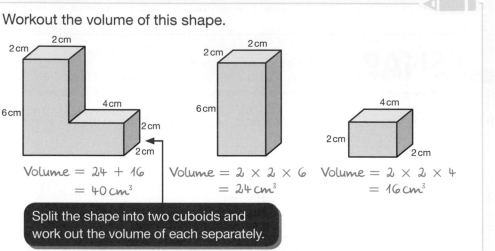

Volume = 24 + 16
        = 40 cm³

Volume = 2 × 2 × 6
        = 24 cm³

Volume = 2 × 2 × 4
        = 16 cm³

Split the shape into two cuboids and work out the volume of each separately.

CHECK**IT!**

1 Calculate the volume of a sphere with radius 4.5 cm. Give your answer to 3 significant figures.

2 A water tank is in the shape of a cylinder on top of a cone. The cylinder is 2 m tall and the whole tank is 3.5 m tall.

If the cylinder has radius 0.5 m, work out the volume of the tank in m³.

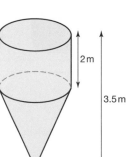

3 The volume of this cone is $k\pi$. Work out the value of k.

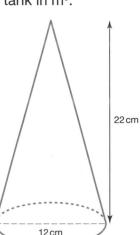

22 cm

12 cm

4 The diagram shows a container partly filled with water to a height of 18 cm.

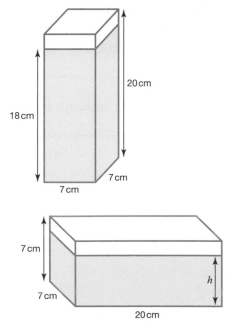

The container is tipped on its side. What is the height of liquid when the container is on its side?

Surface area

The surface area of a three-dimensional shape is the total area of its faces.

Imagine you have to paint the shape – the surface area is the total area you have to paint.

Calculating surface area can be made simpler by considering faces that have the same area.

For example, this cuboid block of wood has pairs of faces with the same area.

Area of top = area of bottom = $9 \times 3 = 27 \, cm^2$

Area of ends = $3 \times 4 = 12 \, cm^2$

Area of front = area of back = $9 \times 4 = 36 \, cm^2$

Surface area = $(2 \times 27) + (2 \times 12) + (2 \times 36)$

$= 54 + 24 + 72$

$= 150 \, cm^2$

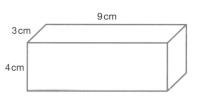

NAILIT!

Surface area is measured in units squared, since you are calculating area.

DOIT!

Find a formula for calculating the surface area of a cuboid measuring a by b by c units.

NAILIT!

Don't just calculate the area of the faces you can see – you need to think about **all** the faces.

You could think about what the net of a shape looks like to help calculate the surface area.

Consider the net of a square-based pyramid.

To calculate the area, first calculate the area of one of the triangular faces and the base.

Surface area = 4 × area of triangle + area of base

WORKIT!

A glass greenhouse is in the shape of a square-based pyramid with its apex directly above the centre of the base.

What area of glass is used?

Area of triangular side $= \frac{1}{2} \times 4 \times 8$

$= 16 \, m^2$

Area of glass = 4×16 ← You do not need to add the area of the base in this instance.

$= 64 \, m^2$

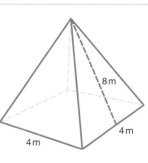

SNAPIT!

Surface areas of shapes

Surface area of a sphere = $4\pi r^2$

Area of curved surface of a cone = $\pi r l$

Remember to use the radius when calculating the surface area of a sphere or cone.

STRETCHIT!

A regular (isosceles) frustum made from a square-based pyramid has parallel top and bottom faces that are square, and four sloping sides that are identical trapeziums.

STRETCHIT!

If you had to find the surface area of a regular frustum, as shown below on the left, you would be given some lengths. You would do the steps below.

1. Find the area of the top and the bottom.
2. Find the area of one side.
3. Add together the areas of top, bottom and 4 x the side.

If your regular frustum is made from a cone, think of it as a large cone with a smaller one cut off the top.

1. Work out both curved surfaces.
2. Take the smaller curved surface away from the larger.
3. Add on the bases of both cones (the top and bottom of the frustum).

WORKIT!

The base of a solid cone has diameter 4 cm and sloping height 8 cm.

Calculate the exact surface area.

Area of sloping face = $\pi \times 2 \times 8$ Diameter = 4, radius = 2.

$\qquad\qquad\qquad = 16\pi$

Area of base = $\pi \times 2^2$ Don't forget to include the area of the base.

$\qquad\qquad = 4\pi$

Total surface area = $16\pi + 4\pi$

$\qquad\qquad\qquad\quad = 20\pi \text{ cm}^2$ If you are asked for the exact solution, leave your answer in terms of π.

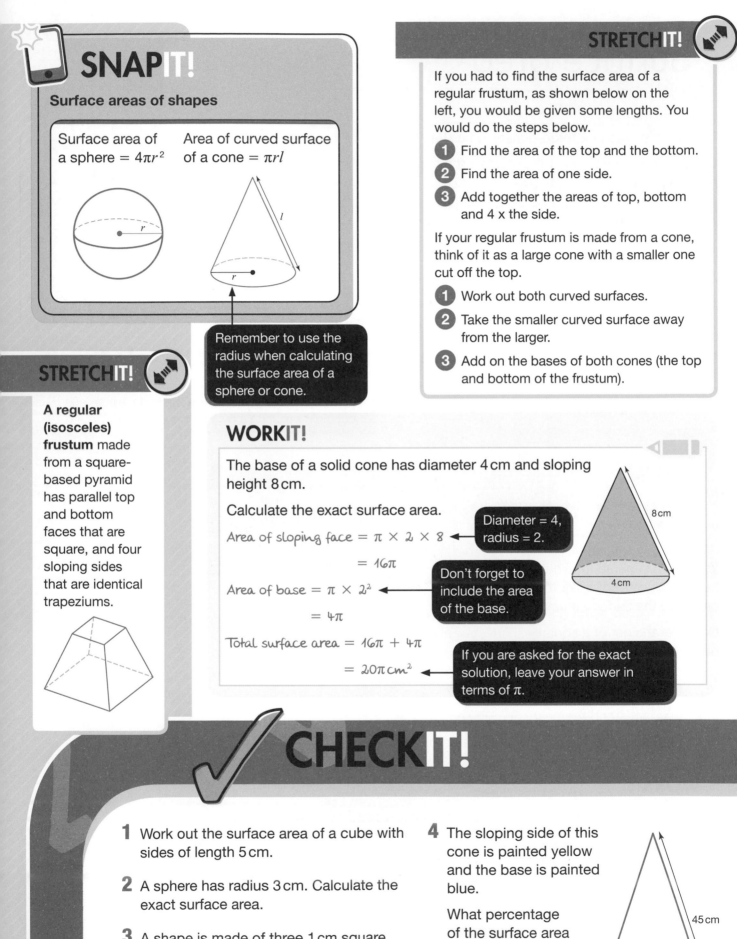

CHECKIT!

1. Work out the surface area of a cube with sides of length 5 cm.

2. A sphere has radius 3 cm. Calculate the exact surface area.

3. A shape is made of three 1 cm square cubes which are glued together and painted.

   Work out the surface area that is painted.

4. The sloping side of this cone is painted yellow and the base is painted blue.

   What percentage of the surface area is yellow?

45 cm

28 cm

Using Pythagoras' theorem

You can use Pythagoras' theorem to find the length of a missing side in a right-angled triangle.

To use Pythagoras' theorem, you need to know the lengths of two sides.

WORKIT!

ABC is a right-angled triangle.

Work out the length of AC.

Give your answer correct to 2 decimal places.

$c^2 = a^2 + b^2$

$c^2 = 10^2 + 9^2$ ◄ Substitute the values of a and b into the formula.

$\quad = 100 + 81 = 181$

$c = \sqrt{181} = 13.4536...$ ◄ Solve, using a calculator to find the square root.

$AC = 13.45\,cm\ (2\ d.p.)$

Label the hypotenuse c. The hypotenuse is the longest side and is opposite the right angle. Label the other two sides a and b.

Round your answer to 2 decimal places and write units with your answer.

The sides could be given as decimal lengths and a range of units (such as mm, cm, m and km) could be used. You also need to be able to give your answer in surd form.

WORKIT!

PQR is a right-angled triangle.

Work out the length of QR.

Give your answer in surd form.

$c^2 = a^2 + b^2$

$6^2 = 3^2 + b^2$ ◄ Substitute the values of a and c into the formula.

$b^2 = 6^2 - 3^2$ ◄ Rearrange and solve.

$\quad = 36 - 9 = 27$

$b = QR = \sqrt{27} = \sqrt{(9 \times 3)}$

$QR = 3\sqrt{3}\ cm^2$

You need to find one of the shorter sides, b.

You can show that a triangle is right-angled by showing that $c^2 = a^2 + b^2$.

NAILIT!

If you are asked to 'give your answer in surd form' **don't** work out the square root.

DOIT!

Show that this triangle is right-angled.

157

Solving problems using Pythagoras' theorem

Problems that require the use of Pythagoras' theorem may involve ladders resting against walls, ramps, flagpoles or 2D shapes that contain a right-angled triangle.

WORKIT!

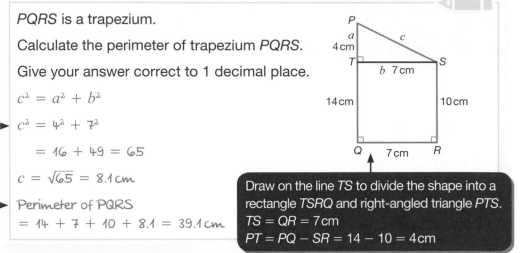

PQRS is a trapezium.

Calculate the perimeter of trapezium *PQRS*.

Give your answer correct to 1 decimal place.

$c^2 = a^2 + b^2$

$c^2 = 4^2 + 7^2$

$\quad = 16 + 49 = 65$

$c = \sqrt{65} = 8.1\,cm$

Perimeter of PQRS

$= 14 + 7 + 10 + 8.1 = 39.1\,cm$

> Substitute the values of a and b into the formula for Pythagoras' theorem. Solve for c.

> Use the value of c (*PS*) to work out the perimeter of *PQRS*.

> Draw on the line *TS* to divide the shape into a rectangle *TSRQ* and right-angled triangle *PTS*.
> $TS = QR = 7\,cm$
> $PT = PQ - SR = 14 - 10 = 4\,cm$

Length of a line segment

A **line segment** is part of a straight line.

You can use Pythagoras' theorem to find the length of a line segment on a coordinate grid.

WORKIT!

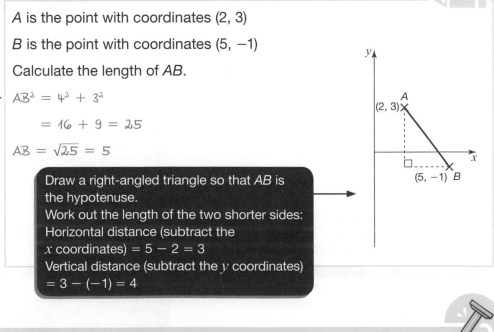

A is the point with coordinates (2, 3)

B is the point with coordinates (5, −1)

Calculate the length of *AB*.

$AB^2 = 4^2 + 3^2$

$\quad = 16 + 9 = 25$

$AB = \sqrt{25} = 5$

> Use Pythagoras' theorem to find the length of *AB*.

> Draw a right-angled triangle so that *AB* is the hypotenuse.
> Work out the length of the two shorter sides:
> Horizontal distance (subtract the x coordinates) $= 5 - 2 = 3$
> Vertical distance (subtract the y coordinates) $= 3 - (-1) = 4$

NAILIT!

If you're not given a diagram as part of the question, you should first draw a sketch of the two points on a set of x- and y-axes.

CHECK IT!

1 *ABC* is a right-angled triangle.

Work out the length of *BC*.

Give your answer correct to 3 significant figures.

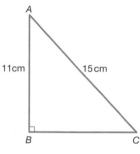

2 A ladder is positioned against a vertical wall at a height of 5.5 m.

The foot of the ladder is 2.4 m from the base of the wall.

How long is the ladder? Give your answer in whole metres.

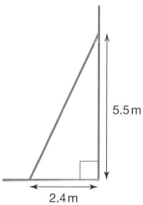

3 *XYZ* is a right-angled triangle.

Calculate the area of the triangle *XYZ*.

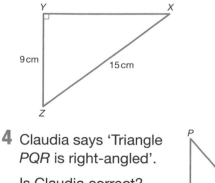

4 Claudia says 'Triangle *PQR* is right-angled'.

Is Claudia correct? Show working to explain your answer.

5 *P* is the point with coordinates (−2, 5)

Q is the point with coordinates (5, −3)

Calculate the length of *PQ*.

Give your answer correct to 2 decimal places.

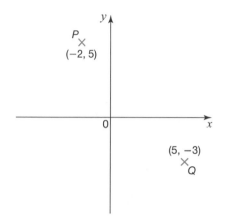

6 *A* has coordinates (2, 4)

B has coordinates (5, 0)

Find the length of the line segment *AB*.

Give your answer in surd form.

7 A field *ABCD* is in the shape of a trapezium. Jesy wants to fence the field. Fencing costs £14 per metre and must be bought in 1 m lengths. Find the cost of the fencing.

8 *ABCD* is a rectangle.

E is the centre of the rectangle.

Work out the length *DE*.

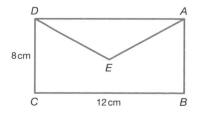

159

Trigonometry

Trigonometric ratios are used to calculate the lengths of sides and angles in right-angled triangles.

In order to use the trigonometric ratios you must label the sides of the triangle.

The **hypotenuse** is the side opposite the right angle – it is the longest side.

The **opposite side** is the one opposite the angle you are working with.

The **adjacent side** is the one next to the angle you are working with.

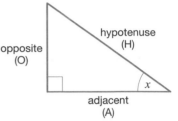

SNAP IT! Trigonometric formulae

$$\sin x = \frac{\text{opposite}}{\text{hypotenuse}} \qquad \cos x = \frac{\text{adjacent}}{\text{hypotenuse}} \qquad \tan x = \frac{\text{opposite}}{\text{adjacent}}$$

Calculating the length of missing sides

Here's how to find the length of side AC in the triangle shown.

1 Label the sides.

2 Write out SOH CAH TOA and underline the sides you know and the one you want to find out.

S<u>O</u>H CAH T<u>O</u>A

3 Use the formula that has **two** letters underlined.

T<u>O</u>A

4 Write out the formula and complete the values you know.

$$\tan 52° = \frac{12}{adjacent}$$

5 Solve the equation – you may need to rearrange it in order to find the missing length.

$$adjacent = \frac{12}{\tan 52°}$$

$$adjacent = 9.4\,\text{cm}$$

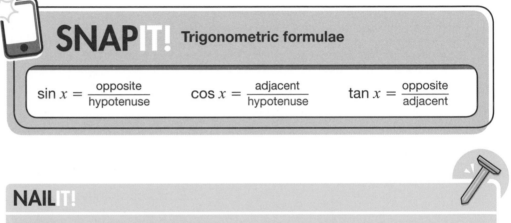

Calculating a missing angle

Here's now to find the size of angle x in this triangle.

1. Label the sides.

2. Write out SOH CAH TOA and underline the sides you know.

   S<u>O</u>H CA<u>H</u> T<u>O</u>A

3. Use the formula that has **two** letters underlined.

   S<u>O</u>H

4. Write out the formula and complete the values you know.

   $\sin x = \dfrac{5}{15}$

5. Find the inverse trigonometric function of your fraction.

   $x = \sin^{-1}\left(\dfrac{5}{15}\right)$

   $x = 19.5°$

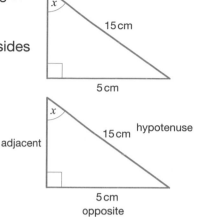

DO IT!

Draw a flow diagram to explain how to find a missing side or angle in a right-angled triangle.

NAIL IT!

The inverse trigonometric functions are $\sin^{-1}$, $\cos^{-1}$ and $\tan^{-1}$, which 'undo' the trigonometric functions sin, cos and tan.

Make sure you can find the inverse trigonometric functions on your calculator – usually you have to press 'shift' and then the trigonometric function.

These are actually the **reciprocals** of the trigonometric functions (see page 33).

Angles of elevation and depression

An angle measured upwards from horizontal is called an angle of elevation.

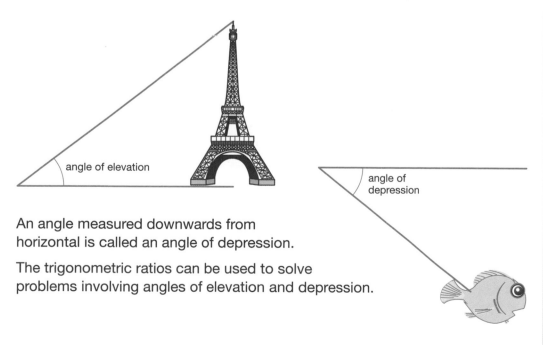

angle of elevation

angle of depression

An angle measured downwards from horizontal is called an angle of depression.

The trigonometric ratios can be used to solve problems involving angles of elevation and depression.

WORKIT!

Write out
SOH CAH TOA and
underline the sides
you know and wish to
know, to identify which
trigonometric formula
to use.

The angle of elevation to a hovering helicopter
is 47° and the vertical height of the helicopter
from the ground is 100 m. How far away
horizontally is the person viewing the helicopter?

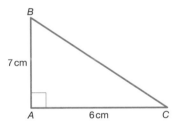

SOH CAH TOA

$$\tan 47° = \frac{100}{\text{horizontal distance}}$$

$$\text{horizontal distance} = \frac{100}{\tan 47}$$

$$= 93.3\,\text{m}$$

Always draw a picture including all
angles and measurements you know.

100 m
opposite

47°

distance =
adjacent

CHECKIT!

1 Use your calculator to work out each of
these values to 1 decimal place.

a tan 23°

b sin 35°

c cos 18°

d $\tan^{-1} 0.5$

e $\sin^{-1}\frac{3}{4}$

f $\cos^{-1}\frac{1}{\sqrt{3}}$

2 Calculate the length of side *MN*. Give
your answer to 2 significant figures.

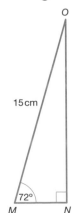

O

15 cm

72°

M   N

3 Calculate the size of angle *ABC*. Give
your answer to 1 decimal place.

B

7 cm

A   6 cm   C

4 An anchor line from a boat is at 15°
to the horizontal. The anchor line is
10 m long.

How deep is the water below the boat?

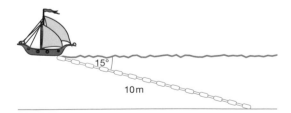

15°

10 m

Exact trigonometric values

There are some special trigonometric values which you will need to learn so you can use them in the non-calculator paper.

There are two special triangles that give you the exact trigonometric values.

Triangle 1

Using SOH CAH TOA and considering the 60° angle gives:

$$\sin 60° = \frac{\sqrt{3}}{2} \qquad \cos 60° = \frac{1}{2} \qquad \tan 60° = \sqrt{3}$$

Considering the 30° angle gives:

$$\sin 30° = \frac{1}{2} \qquad \cos 30° = \frac{\sqrt{3}}{2} \qquad \tan 30° = \frac{1}{\sqrt{3}}$$

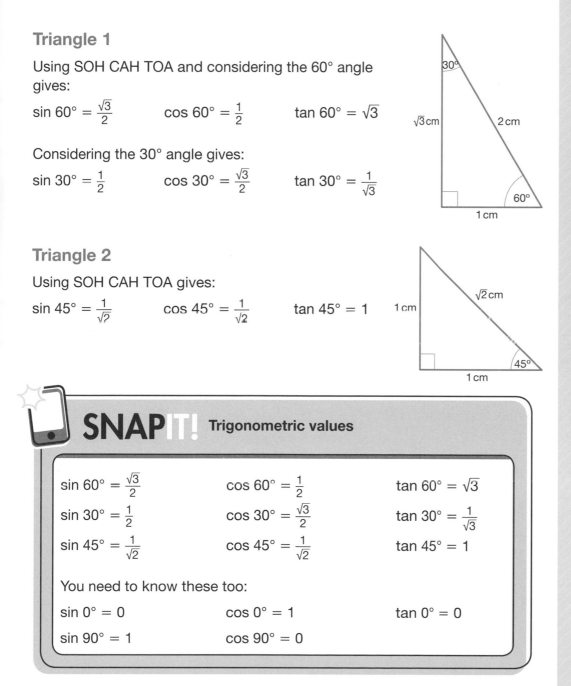

Triangle 2

Using SOH CAH TOA gives:

$$\sin 45° = \frac{1}{\sqrt{2}} \qquad \cos 45° = \frac{1}{\sqrt{2}} \qquad \tan 45° = 1$$

SNAPIT! **Trigonometric values**

$\sin 60° = \frac{\sqrt{3}}{2}$	$\cos 60° = \frac{1}{2}$	$\tan 60° = \sqrt{3}$
$\sin 30° = \frac{1}{2}$	$\cos 30° = \frac{\sqrt{3}}{2}$	$\tan 30° = \frac{1}{\sqrt{3}}$
$\sin 45° = \frac{1}{\sqrt{2}}$	$\cos 45° = \frac{1}{\sqrt{2}}$	$\tan 45° = 1$

You need to know these too:

$\sin 0° = 0$	$\cos 0° = 1$	$\tan 0° = 0$
$\sin 90° = 1$	$\cos 90° = 0$	

You can also write these as ratios.

For example, $\sin 60° = \frac{\sqrt{3}}{2}$ tells you that in a right-angled triangle with an angle of 60°, the ratio between the opposite side and the hypotenuse is $\sqrt{3} : 2$.

DOIT!

1 Write a ratio fact for each of the trigonometric values in the Snap it!

2 Make a set of cards. On one side write the sin/cos/tan values of 0, 30, 45, 60 or 90.

On the other, write the correct ratio.

WORKIT!

Triangle *ABC* has a right angle at *B*.

Angle *BAC* = 60° and side *AC* = 10 cm.

Work out the length of side *AB*.

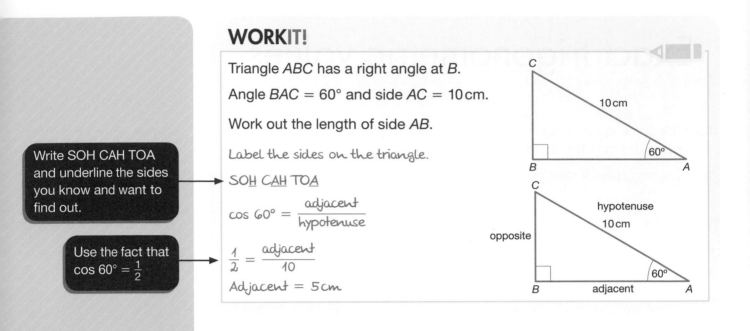

Write SOH CAH TOA and underline the sides you know and want to find out.

Label the sides on the triangle.

SO<u>H</u> C<u>A</u><u>H</u> TO<u>A</u>

$$\cos 60° = \frac{adjacent}{hypotenuse}$$

Use the fact that $\cos 60° = \frac{1}{2}$

$$\frac{1}{2} = \frac{adjacent}{10}$$

Adjacent = 5 cm

CHECKIT!

Complete these questions without using a calculator.

1 Write down the value of each of these.

   a sin 30°   **d** sin 45°

   b cos 90°   **e** tan 60°

   c tan 0°

2 Work out the lengths of all the missing sides in this triangle:

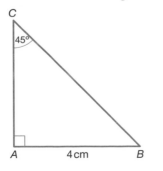

3 In a right-angled triangle, the ratio between the opposite and adjacent sides is 1 : √3. What are the sizes of the other angles in the triangle?

4 Write down the size of angle *ABC* in the triangle below. Explain how you know.

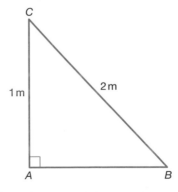

5 Put these values in order of size from smallest to largest:

   0.5   cos 30° $\frac{3}{4}$ tan 45°

Transformations

Translations

A **translation** is a sliding movement. All the points on a 2D shape move the same distance in the same direction.

To describe a translation you use a **column vector**.

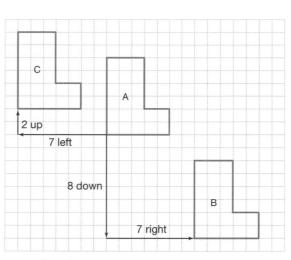

The transformation that moves shape A to shape B is a translation by the vector $\begin{pmatrix} 7 \\ -8 \end{pmatrix}$

The top number gives the movement parallel to the horizontal axis (right or left).

The bottom number gives the movement parallel to the vertical axis (up or down).

The transformation that moves shape A to shape C is a translation by the vector $\begin{pmatrix} -7 \\ 2 \end{pmatrix}$

A negative top number means the movement is to the left. A positive bottom number means the movement is up.

A positive top number means the movement is to the right. A negative bottom number means the movement is down.

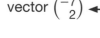

NAILIT!

To describe a translation you need to give the column vector.

WORKIT!

On the grid, translate triangle P by the vector $\begin{pmatrix} 5 \\ -2 \end{pmatrix}$

Label the new shape Q.

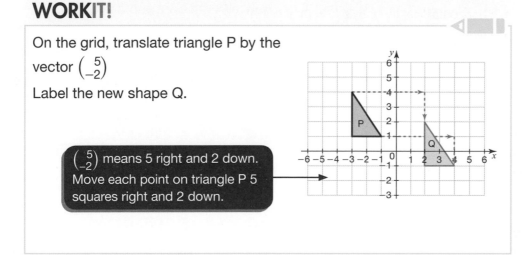

$\begin{pmatrix} 5 \\ -2 \end{pmatrix}$ means 5 right and 2 down. Move each point on triangle P 5 squares right and 2 down.

DOIT!

Copy the grid and triangle P from Work it! Translate triangle P by the vector

a $\begin{pmatrix} 2 \\ 2 \end{pmatrix}$ **c** $\begin{pmatrix} 0 \\ 3 \end{pmatrix}$

b $\begin{pmatrix} -2 \\ -2 \end{pmatrix}$ **d** $\begin{pmatrix} 3 \\ 0 \end{pmatrix}$

The vector to translate shape A to shape B is $\begin{pmatrix} 5 \\ -1 \end{pmatrix}$

Write down the vector that translates shape B to shape A.

Translated shapes are **congruent** (see page 135). The lengths of the sides of the shape do not change and the angles of the shape do not change.

Reflections

A **reflection** is obtained when a 2D shape is reflected in a mirror line.

Every point on the **object** (the original 2D shape) and the **image** (the shape following the transformation) must be the same distance away from the mirror line.

You can reflect a shape on a squared grid.

Reflected shapes are congruent.

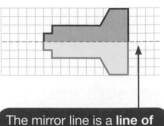

> The mirror line is a **line of symmetry** (see page 133).

WORKIT!

> The question says 'describe fully' so you need to write down the type of transformation: a reflection and the equation of the mirror line: $y = x$.

Describe fully the transformation that maps triangle A onto triangle B.

A reflection in the line $y = x$.

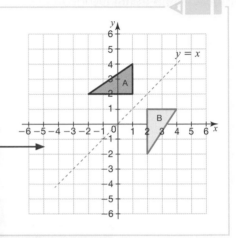

> Draw on the line of symmetry. You can use tracing paper to check that you have identified the correct mirror line. Trace shape A and the mirror line. Remember to mark a reference point on the mirror line, like the origin. Turn the tracing paper over and line up the mirror lines. Does shape A map onto shape B?

NAILIT!

To describe a reflection you need to give the equation of the mirror line.

The mirror line could be:

- a horizontal line: the x-axis ($y = 0$), $y = 1$, $y = 2$...
- a vertical line: the y-axis ($x = 0$), $x = -1$, $x = 1$...
- a diagonal line: $y = x$, $y = -x$

(See page 76 if you're unsure about the equations of straight lines.)

WORKIT!

On the grid, reflect shape P in the line $x = 1$.

Label the new shape Q.

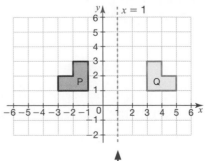

> Draw on the line of symmetry, $x = 1$.

DOIT!

Copy the grid and shape P from Work it! Reflect shape P in

a the y-axis

b the x-axis

c the line $y = 1$

Rotations

A **rotation** can be described as a fraction of a turn or as an angle of turn.

The point about which the 2D shape is turned is called the **centre of rotation**.

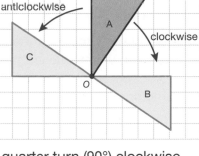

Using O as the centre of rotation:

- the rotation of triangle A to triangle B is one quarter turn (90°) clockwise
- the rotation of triangle A to triangle C is one quarter turn (90°) anticlockwise
- the rotation of triangle B to triangle C is one half turn (180°). ◀——

> When the angle of turn is 180° you don't need to give a direction.

Rotated shapes are congruent.

WORKIT!

On the grid, rotate shape A 90° clockwise about the origin, O.

Label the new shape B.

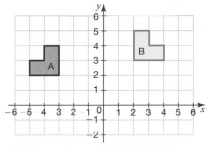

> Mark the centre of rotation with an ×.
> Trace shape A. Put your pencil on the ×.
> Rotate the tracing paper 90° ($\frac{1}{4}$ turn) clockwise.

You can also use tracing paper to help you find the centre of rotation.

WORKIT!

Describe fully the single transformation that maps triangle S onto triangle T.

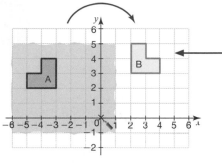

Rotation of 90° anticlockwise about (−1, 1).

> From triangle S to triangle T, the direction is anticlockwise and the turn is 90°.
> Draw lines to find the centre of rotation.
> Or, trace shape S. Put your pencil on a point and rotate the tracing paper. Repeat with a different point until triangle S maps onto triangle T.

DOIT!

Copy the grid and shape A from Work it! Rotate shape A
a 180° about the point O
b 90° clockwise about (1, 0)
c 90° anticlockwise about (0, 2)

NAILIT!

To describe a rotation you need to give:
- the angle of rotation
- the direction of rotation (clockwise or anticlockwise)
- the centre of rotation.

The centre of rotation could be the origin O or any other point on a coordinate grid.

Enlargement

An **enlargement** changes the size of an object. All the side lengths of a 2D shape are multiplied by the same **scale factor**.

Here, shape A has been enlarged by scale factor 2.

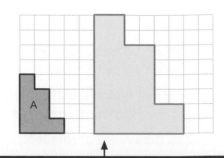

Enlarged shapes are **not** congruent as the lengths of the sides have changed. Enlarged shapes are **similar** (see page 171).

> Each length on the enlarged shape is 2 times the corresponding length on shape A.

When you are given a **centre of enlargement**, you need to multiply the distance from the centre to each point on the shape by the scale factor.

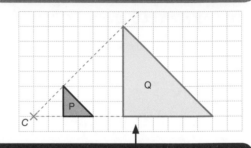

> For a scale factor of 3, each point on triangle Q is three times as far from the centre C as the corresponding point on triangle P.

DOIT!

Copy the grid and triangle A from Work it! Enlarge triangle A by

a scale factor 2 about centre (1, 1)

b scale factor $1\frac{1}{2}$ about centre (1, 5)

c scale factor 2 about centre (3, 2).

WORKIT!

On the grid, enlarge triangle A by scale factor $\frac{1}{2}$ about centre (1, 5).

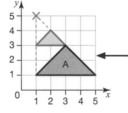

> As the scale factor is less than 1 the **image** will be smaller than the **object** (triangle A).
> For a scale factor of $\frac{1}{2}$, multiply all the distances from the centre by $\frac{1}{2}$ (or divide all the distances by 2).

NAILIT!

To describe an enlargement you need to give

• the scale factor **and**

• the centre of enlargement.

WORKIT!

Describe fully the single transformation that maps shape A onto shape B.

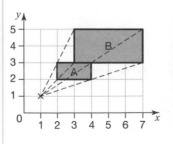

> First work out the scale factor.
> Scale factor = $\frac{\text{enlarged length}}{\text{original length}} = \frac{4}{2} = 2$
> To find the centre of enlargement, draw lines through corresponding points on the object and image. Extend the lines until they meet at the centre of enlargement.

Enlargement by scale factor 2, centre (1, 1).

Combinations of transformations

You may need to find a single transformation that describes two (or more) separate transformations.

A single translation can be used to describe two or more translations.

A single rotation can be used to describe a reflection in the y-axis followed by a reflection in the x-axis.

Shape A to B is a translation by $\begin{pmatrix} 2 \\ -5 \end{pmatrix}$
Shape B to C is a translation by $\begin{pmatrix} 6 \\ 3 \end{pmatrix}$
Shape A to C is a translation by $\begin{pmatrix} 8 \\ -2 \end{pmatrix}$

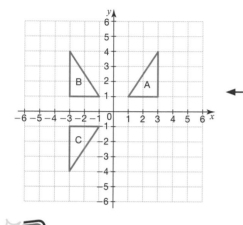

A to B is a reflection in the y-axis.
B to C is a reflection in the x-axis.
A to C is a rotation of 180° about O.

STRETCH IT!

Can a reflection in one axis followed by a reflection in the other axis always be described by a rotation?

SNAP IT! Describing transformations

To describe fully:

- **a translation**, give the column vector

- **a reflection**, give the equation of the mirror line

- **a rotation**, give the angle of rotation, the direction of rotation, and the centre of rotation

- **an enlargement**, give the scale factor and the centre of enlargement.

WORK IT!

a On the grid, reflect triangle P in the x-axis. Label the image Q.

b Reflect triangle Q in the line $x = 1$. Label the image R.

c Describe fully the single transformation that maps triangle P onto triangle R.

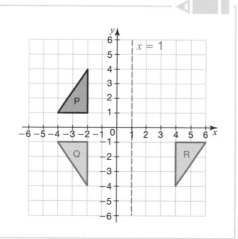

Rotation of 180° about centre (1, 0).

CHECK IT!

1 On a copy of the grid, draw an enlargement, scale factor 2, of the shaded shape.

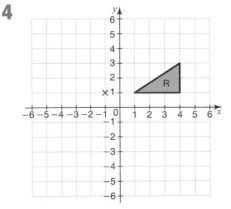

2 On a copy of the grid, enlarge triangle A by scale factor $\frac{1}{2}$ about centre P.

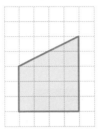

3 Describe fully the single transformation that maps shape A onto shape B.

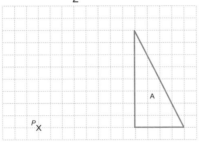

4

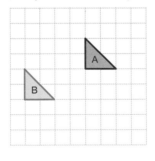

a Rotate triangle R a half turn about the point O. Label the new shape S.

b Rotate triangle R 90° anticlockwise about the point $(-1, 1)$. Label the new shape T.

5 Describe fully the single transformation that maps shape P onto shape Q.

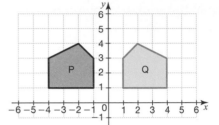

6 Describe fully the single transformation that maps shape A onto shape B.

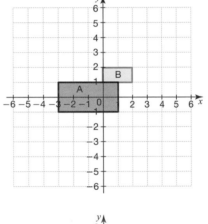

7

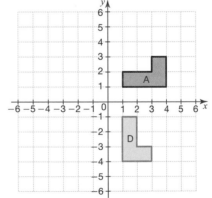

a Reflect shape A in the y-axis. Label the new shape B.

b Translate shape A by vector $\begin{pmatrix} -2 \\ -3 \end{pmatrix}$. Label the new shape C.

c Describe fully the single transformation that maps shape A onto shape D.

Similar shapes

When one shape is an enlargement (see page 168) of another, the shapes are **similar**.

In similar shapes, the corresponding angles are the same and the corresponding sides are in the same ratio. This ratio is the **scale factor** of the enlargement.

Triangles *ABC* and *DEF* are similar.

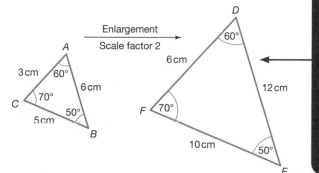

The three pairs of corresponding angles are:
ACB and *DFE*
CAB and *FDE*
ABC and *DEF*.

The three pairs of corresponding sides are:
AC and *DF*
BC and *EF*
AB and *DE*.

STRETCH**IT!**

Work out the perimeter of triangle *ABC* and the perimeter of triangle *DEF*.
When a shape is enlarged by scale factor 2, by what scale factor is the perimeter of the shape enlarged?
Write down a general statement about the perimeter of an enlarged shape.

WORK**IT!**

Triangle *RST* is similar to triangle *WUV*.

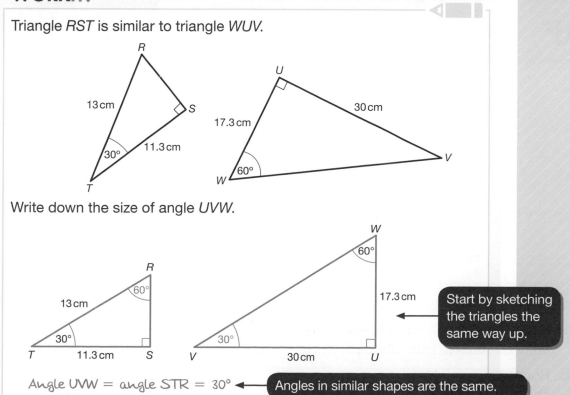

Write down the size of angle *UVW*.

Start by sketching the triangles the same way up.

Angle UVW = angle STR = 30°   Angles in similar shapes are the same.

171

To work out a corresponding length you first need to work out the scale factor.

Here are two similar triangles, *GHI* and *JKL*.

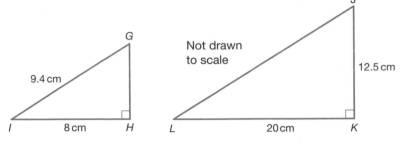

Not drawn to scale

Take two corresponding sides and work out the scale factor as shown below.

Look at two corresponding sides where you know the lengths of both: *HI* (8 cm) and *KL* (20 cm).

Scale factor of enlargement $= \dfrac{\text{enlarged length}}{\text{original length}} = \dfrac{20}{8} = \dfrac{10}{4} = 2.5$

Once you know the scale factor, you can use this to work out unknown lengths such as *JL* or *GH*.

Side *GH* corresponds to side *JK*.

The length of *GH* will be shorter than the length of *JK* so divide by the scale factor.

Length of $JL = 9.4 \text{ cm} \times 2.5 = 23.5 \text{ cm}$

Length of $GH = 12.5 \text{ cm} \div 2.5 = 5 \text{ cm}$

Side *JL* corresponds to side *GI*. The length of *JL* will be longer than the length of *GI* so multiply by the scale factor.

To show that two triangles are similar you need to show that one of the following conditions is satisfied.

- All three pairs of angles are equal.

- All three pairs of sides are in the same ratio.

- Two pairs of sides, with the same angle between them, are in the same ratio.

Here, the scale factor of enlargement is 3.

$$\dfrac{DE}{AB} = \dfrac{FE}{CB} = \dfrac{DF}{AC}$$

$$\dfrac{15}{5} = \dfrac{12}{4} = \dfrac{9}{3} = 3$$

You can write:

The ratio $AB : DE = 1 : 3$

or:

The ratio $FE : CB = 3 : 1$

SNAP IT! Identifying similar triangles

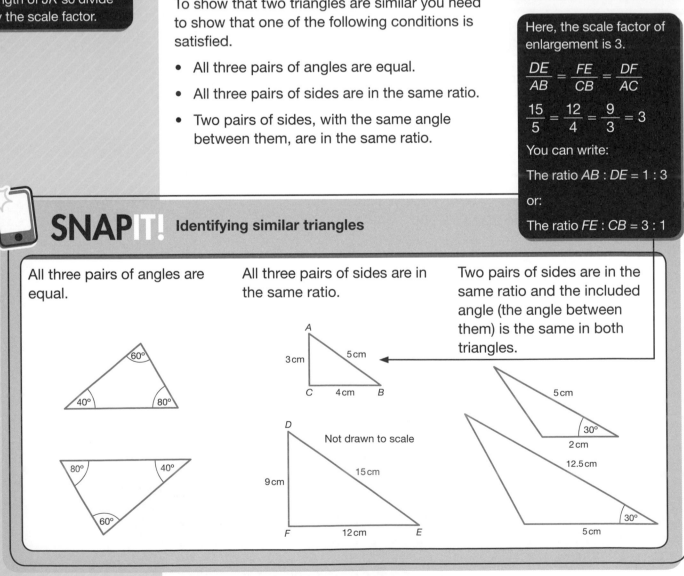

All three pairs of angles are equal.

All three pairs of sides are in the same ratio.

Two pairs of sides are in the same ratio and the included angle (the angle between them) is the same in both triangles.

STRETCHIT!

Show that triangles *ABC* and *EDC* are similar.

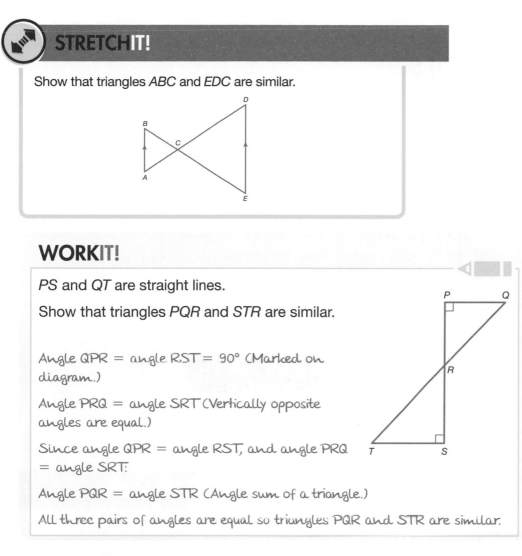

WORKIT!

PS and *QT* are straight lines.

Show that triangles *PQR* and *STR* are similar.

Angle QPR = angle RST = 90° (Marked on diagram.)

Angle PRQ = angle SRT (Vertically opposite angles are equal.)

Since angle QPR = angle RST, and angle PRQ = angle SRT:

Angle PQR = angle STR (Angle sum of a triangle.)

All three pairs of angles are equal so triangles PQR and STR are similar.

DOIT!

In the Work it! length
PQ = 2.5 cm
The ratio *PR* : *RS* = 1 : 4
Find the length *ST*.

CHECKIT!

1 Triangle *ABC* is similar to triangle *DEF*.

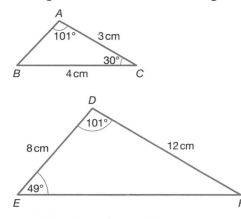

a Write down the size of angle *DFE*.

b Work out the length of *EF*.

c Work out the length of *AB*.

2 Shape *LMNO* is similar to shape *PQRS*.

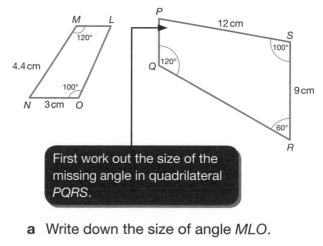

First work out the size of the missing angle in quadrilateral *PQRS*.

a Write down the size of angle *MLO*.

b Work out the length of *QR*.

c Work out the length of *LO*.

Vectors

A vector has **magnitude** (size or length) and direction.

Vectors can be written as a single bold letter such as **a** or **b**, as a **column vector** (see page 165), or with arrows, $\overrightarrow{AB}$ and they can be drawn on squared paper.

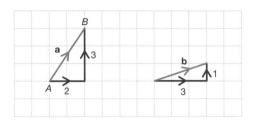

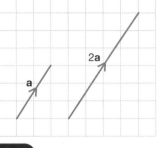

You can multiply a vector by a number.

If $\mathbf{a} = \begin{pmatrix} 2 \\ 3 \end{pmatrix}$ then $2\mathbf{a} = \begin{pmatrix} 4 \\ 6 \end{pmatrix}$ ◀ $2\mathbf{a} = 2 \times \mathbf{a} = \begin{pmatrix} 2 \times 2 \\ 2 \times 3 \end{pmatrix}$

The vectors **a** and 2**a** are in the same direction and are parallel, but 2**a** is twice as long.

For a vector **b**, the vector −**b** has the same length but is in the opposite direction.

If $\mathbf{b} = \begin{pmatrix} 3 \\ 1 \end{pmatrix}$ then $-\mathbf{b} = \begin{pmatrix} -3 \\ -1 \end{pmatrix}$ and $-3\mathbf{b} = \begin{pmatrix} -9 \\ -3 \end{pmatrix}$ ◀ $-3\mathbf{b} = -3 \times \mathbf{b} = \begin{pmatrix} -3 \times 3 \\ -3 \times 1 \end{pmatrix}$

SNAP IT!   Parallel vectors

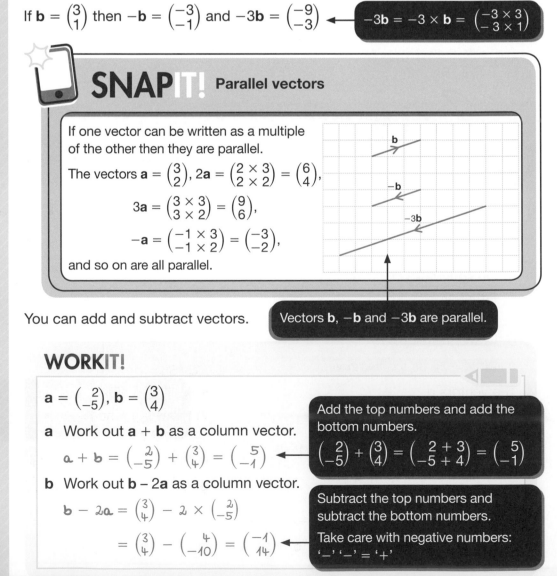

If one vector can be written as a multiple of the other then they are parallel.

The vectors $\mathbf{a} = \begin{pmatrix} 3 \\ 2 \end{pmatrix}$, $2\mathbf{a} = \begin{pmatrix} 2 \times 3 \\ 2 \times 2 \end{pmatrix} = \begin{pmatrix} 6 \\ 4 \end{pmatrix}$,

$3\mathbf{a} = \begin{pmatrix} 3 \times 3 \\ 3 \times 2 \end{pmatrix} = \begin{pmatrix} 9 \\ 6 \end{pmatrix}$,

$-\mathbf{a} = \begin{pmatrix} -1 \times 3 \\ -1 \times 2 \end{pmatrix} = \begin{pmatrix} -3 \\ -2 \end{pmatrix}$,

and so on are all parallel.

You can add and subtract vectors.

Vectors **b**, −**b** and −3**b** are parallel.

WORKIT!

$\mathbf{a} = \begin{pmatrix} 2 \\ -5 \end{pmatrix}$, $\mathbf{b} = \begin{pmatrix} 3 \\ 4 \end{pmatrix}$

a Work out **a** + **b** as a column vector.

$\mathbf{a} + \mathbf{b} = \begin{pmatrix} 2 \\ -5 \end{pmatrix} + \begin{pmatrix} 3 \\ 4 \end{pmatrix} = \begin{pmatrix} 5 \\ -1 \end{pmatrix}$

Add the top numbers and add the bottom numbers.

$\begin{pmatrix} 2 \\ -5 \end{pmatrix} + \begin{pmatrix} 3 \\ 4 \end{pmatrix} = \begin{pmatrix} 2 + 3 \\ -5 + 4 \end{pmatrix} = \begin{pmatrix} 5 \\ -1 \end{pmatrix}$

b Work out **b** − 2**a** as a column vector.

$\mathbf{b} - 2\mathbf{a} = \begin{pmatrix} 3 \\ 4 \end{pmatrix} - 2 \times \begin{pmatrix} 2 \\ -5 \end{pmatrix}$

$= \begin{pmatrix} 3 \\ 4 \end{pmatrix} - \begin{pmatrix} 4 \\ -10 \end{pmatrix} = \begin{pmatrix} -1 \\ 14 \end{pmatrix}$

Subtract the top numbers and subtract the bottom numbers.

Take care with negative numbers: '−' '−' = '+'

NAIL IT!

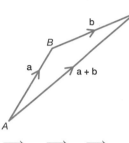

Resultant vector

$\overrightarrow{AC} = \overrightarrow{AB} + \overrightarrow{BC}$

$\quad = \mathbf{a} + \mathbf{b}$

The resultant vector of the vectors **a** and **b** is **a** + **b**.

DO IT!

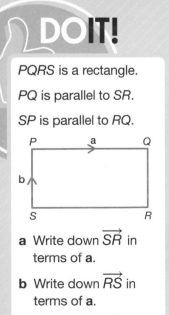

PQRS is a rectangle.

PQ is parallel to SR.

SP is parallel to RQ.

a Write down $\overrightarrow{SR}$ in terms of **a**.

b Write down $\overrightarrow{RS}$ in terms of **a**.

c Write down $\overrightarrow{QR}$ in terms of **b**.

You can also draw vectors to represent the sum and difference of two vectors.

The two vectors **a** and **b**	The sum of the two vectors, **a** + **b**	The difference of the two vectors, **a** − **b**

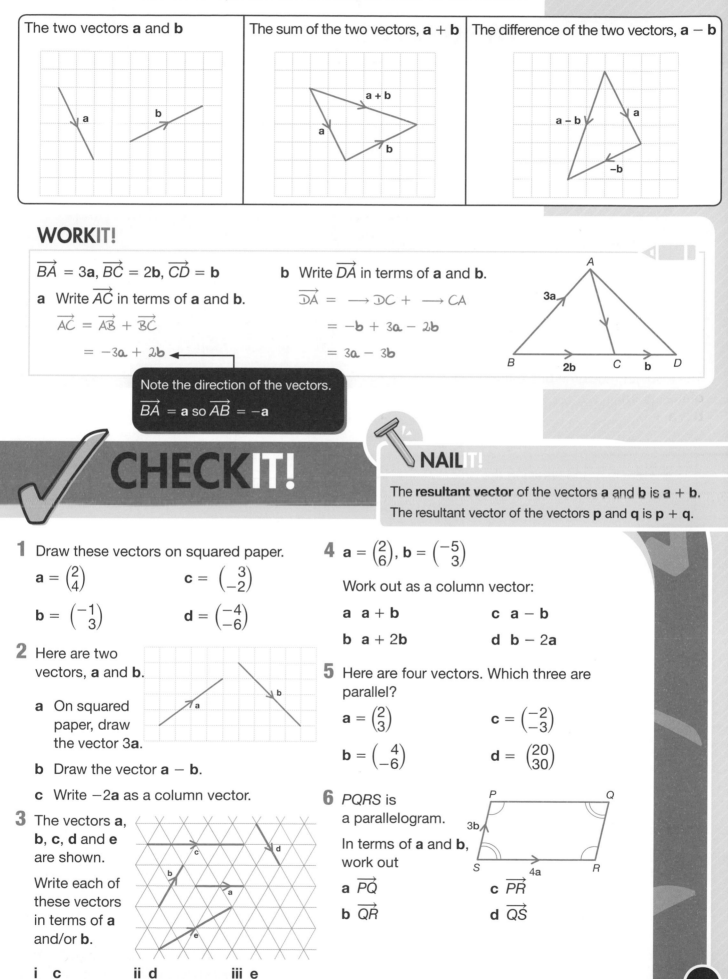

WORKIT!

$\overrightarrow{BA}$ = 3**a**, $\overrightarrow{BC}$ = 2**b**, $\overrightarrow{CD}$ = **b**

a Write $\overrightarrow{AC}$ in terms of **a** and **b**.

$\overrightarrow{AC} = \overrightarrow{AB} + \overrightarrow{BC}$

    = −3**a** + 2**b**

b Write $\overrightarrow{DA}$ in terms of **a** and **b**.

$\overrightarrow{DA}$ = ⟶ $\overrightarrow{DC}$ + ⟶ $\overrightarrow{CA}$

    = −**b** + 3**a** − 2**b**

    = 3**a** − 3**b**

> Note the direction of the vectors.
> $\overrightarrow{BA}$ = **a** so $\overrightarrow{AB}$ = −**a**

✓ CHECKIT!

NAILIT!

The **resultant vector** of the vectors **a** and **b** is **a** + **b**.

The resultant vector of the vectors **p** and **q** is **p** + **q**.

1 Draw these vectors on squared paper.

$\mathbf{a} = \begin{pmatrix} 2 \\ 4 \end{pmatrix}$ $\mathbf{c} = \begin{pmatrix} 3 \\ -2 \end{pmatrix}$

$\mathbf{b} = \begin{pmatrix} -1 \\ 3 \end{pmatrix}$ $\mathbf{d} = \begin{pmatrix} -4 \\ -6 \end{pmatrix}$

2 Here are two vectors, **a** and **b**.

a On squared paper, draw the vector 3**a**.

b Draw the vector **a** − **b**.

c Write −2**a** as a column vector.

3 The vectors **a**, **b**, **c**, **d** and **e** are shown.

Write each of these vectors in terms of **a** and/or **b**.

   i **c**   **ii** **d**   **iii** **e**

4 $\mathbf{a} = \begin{pmatrix} 2 \\ 6 \end{pmatrix}$, $\mathbf{b} = \begin{pmatrix} -5 \\ 3 \end{pmatrix}$

Work out as a column vector:

   a **a** + **b**   **c** **a** − **b**

   b **a** + 2**b**   **d** **b** − 2**a**

5 Here are four vectors. Which three are parallel?

$\mathbf{a} = \begin{pmatrix} 2 \\ 3 \end{pmatrix}$ $\mathbf{c} = \begin{pmatrix} -2 \\ -3 \end{pmatrix}$

$\mathbf{b} = \begin{pmatrix} 4 \\ -6 \end{pmatrix}$ $\mathbf{d} = \begin{pmatrix} 20 \\ 30 \end{pmatrix}$

6 PQRS is a parallelogram.

In terms of **a** and **b**, work out

   a $\overrightarrow{PQ}$ **c** $\overrightarrow{PR}$

   b $\overrightarrow{QR}$ **d** $\overrightarrow{QS}$

You may **not** use a calculator for these questions.

1 Here is trapezium *ABCD*. The diagram is accurately drawn.

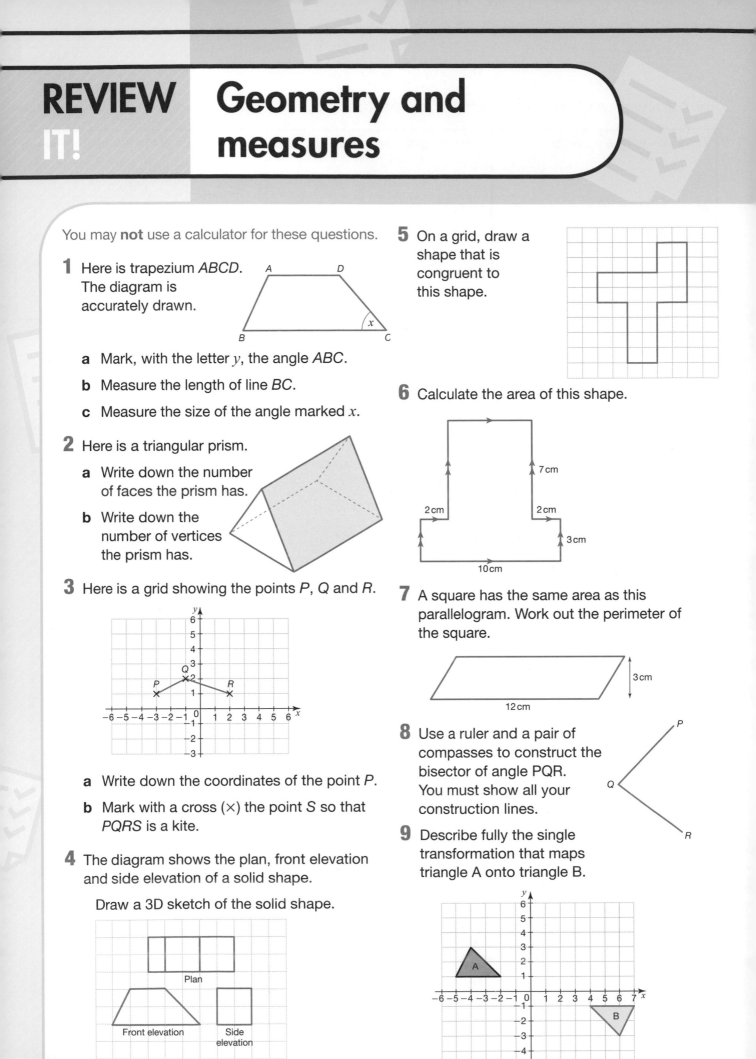

a Mark, with the letter *y*, the angle *ABC*.

b Measure the length of line *BC*.

c Measure the size of the angle marked *x*.

2 Here is a triangular prism.

a Write down the number of faces the prism has.

b Write down the number of vertices the prism has.

3 Here is a grid showing the points *P*, *Q* and *R*.

a Write down the coordinates of the point *P*.

b Mark with a cross (×) the point *S* so that *PQRS* is a kite.

4 The diagram shows the plan, front elevation and side elevation of a solid shape.

Draw a 3D sketch of the solid shape.

Plan

Front elevation   Side elevation

5 On a grid, draw a shape that is congruent to this shape.

6 Calculate the area of this shape.

7 cm

2 cm   2 cm

3 cm

10 cm

7 A square has the same area as this parallelogram. Work out the perimeter of the square.

3 cm

12 cm

8 Use a ruler and a pair of compasses to construct the bisector of angle PQR. You must show all your construction lines.

P

Q

R

9 Describe fully the single transformation that maps triangle A onto triangle B.

10 Here is part of a regular 8-sided polygon.

Work out the size of the angle marked x.

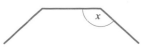

11 *ABD* and *GFE* are parallel.
$GC = GF$.

Work out the size of the angle marked x.
You must show your working.

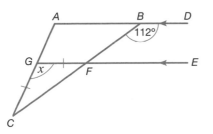

12 What proportion of the rectangle is shaded?

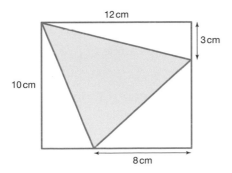

13 In triangle *ABC*, $AB = 6$ cm, $BC = 4$ cm, $AC = 8$ cm.

By calculation, decide whether triangle *ABC* is a right-angled triangle.

14 Using ruler and protractor only, construct a triangle *PQR* such that $PQ = 8$ cm, angle $RPQ = 60°$ and angle $PQR = 45°$.

15 The diagram shows two sides of a quadrilateral.

Scale: 1 cm = 2 m

a Use ruler and compasses to construct the shortest path from point *A* to each side.

b Work out the difference in the distances.

16 a Write down the exact value of cos 45°.

b Given that $\cos 60° = \frac{1}{2}$, work out the length of side *AB*.

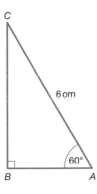

17 $\mathbf{a} = \begin{pmatrix} 4 \\ -5 \end{pmatrix}$, $\mathbf{b} = \begin{pmatrix} 2 \\ 3 \end{pmatrix}$

a Work out $\mathbf{a} - \mathbf{b}$ as a column vector.

b Work out $\mathbf{a} + 2\mathbf{b}$ as a column vector.

c Work out $\mathbf{b} - 2\mathbf{a}$ as a column vector.

You **may** use a calculator for these questions.

18 On squared paper, draw a kite.

19 A shape has three lines of symmetry.

All sides are the same length.

Write down the name of the shape.

20 The diagram shows a solid made from four cubes.

On a grid, draw a plan of the solid.

21

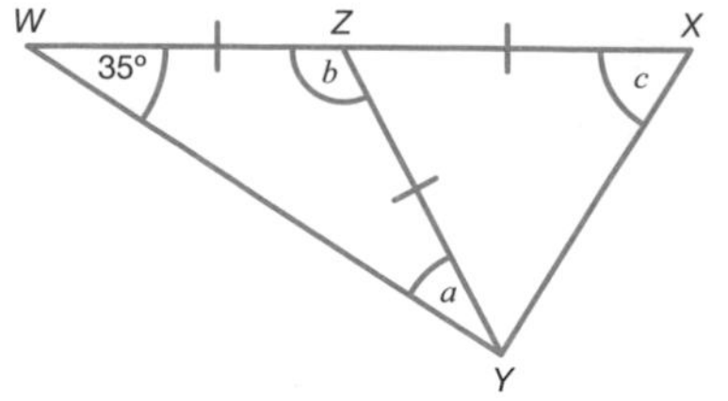

WZX is a straight line.

WZ = XZ = YZ

a i Write down the size of the angle marked a.

ii Give a reason for your answer.

b Work out the size of the angle marked b.

c Work out the size of the angle marked c.

22 ABC is a right-angled triangle.

Work out the length of BC. Give your answer correct to 1 decimal place.

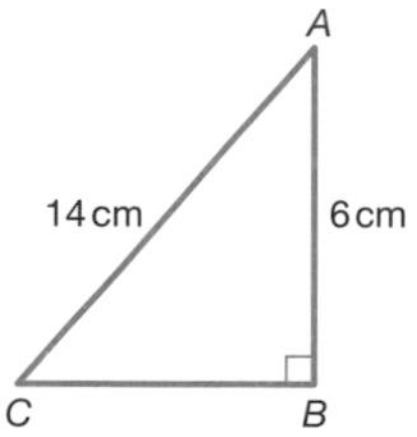

23 The diagram shows part of a pattern made from squares and regular octagons.

Find the size of the angle marked x.

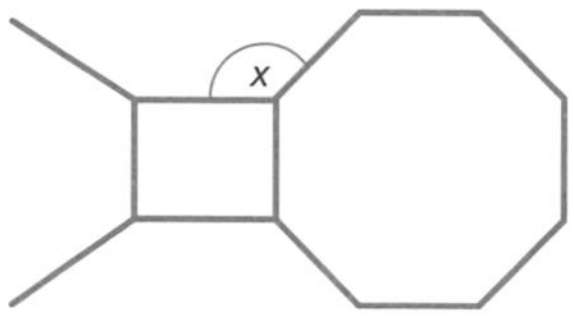

24 PQR is a triangle.

Make an accurate copy of PQR.

Shade the region inside the triangle that is less than 3 cm from the point Q and closer to the line PQ than the line QR.

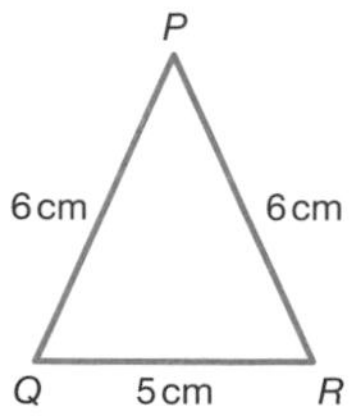

25 ABCD is a trapezium.

Calculate the perimeter of trapezium ABCD. Give your answer correct to 2 decimal places.

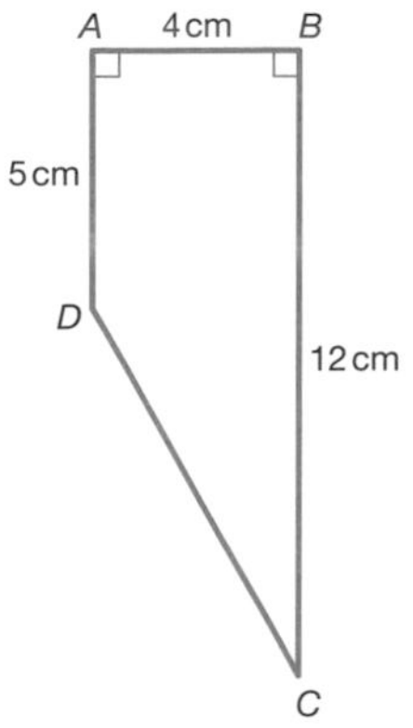

26 The shape below is a rectangle with a quarter-circle on top.

Calculate the perimeter of the shape to 1 decimal place.

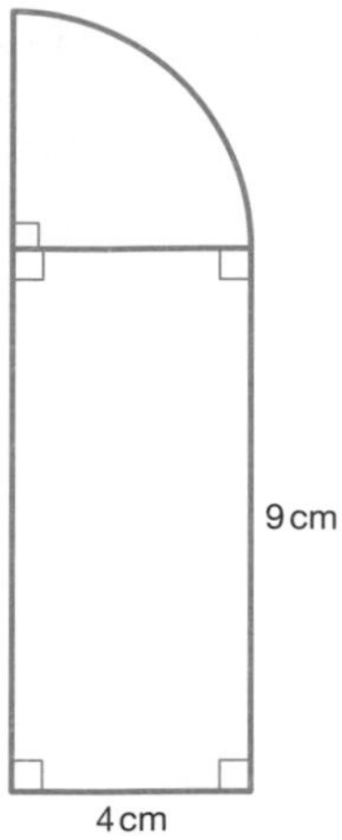

27 The largest circle possible is drawn inside a square with sides 6 cm long.

Work out the area that is shaded. Give your answer correct to 1 decimal place.

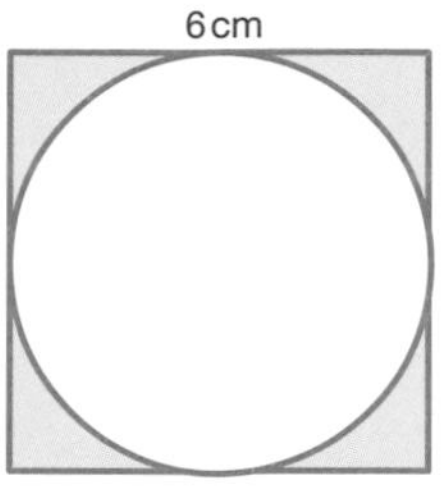

28 The diagram shows a cylindrical glass with height 15 cm and radius 3 cm.

How many times can the glass be completely filled from a 2–litre bottle of water?

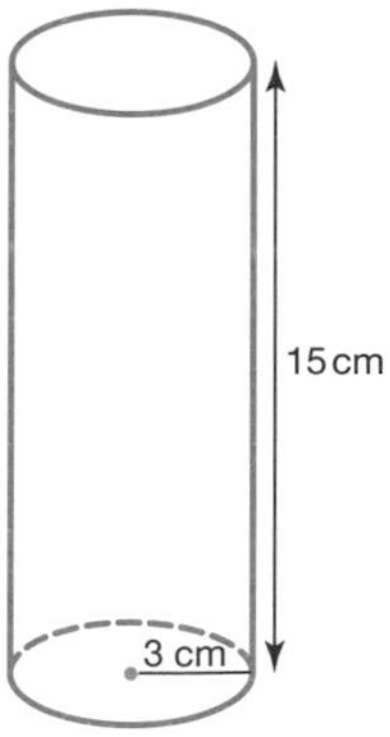

29

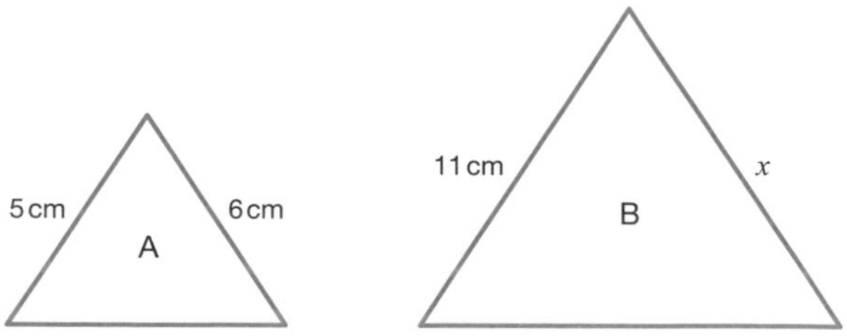

Triangles A and B are similar.

Work out x.

30

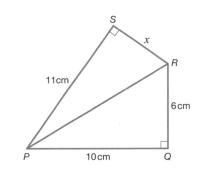

Triangles *PQR* and *RSP* are right-angled triangles.

Work out the value of *x*. Give your answer correct to 3 significant figures.

31 A solid cone has a circular base with diameter 12 cm, sloping height 10 cm and vertical height 8 cm.

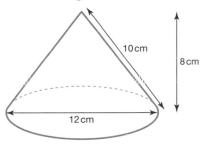

Work out:

a the surface area

b the volume.

Give your answers to 2 significant figures.

32 Calculate the size of angle *x* in the triangle. Give your answer to one decimal place.

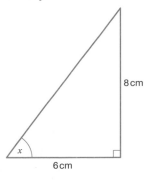

33 Describe fully the single transformation that maps shape A onto shape B.

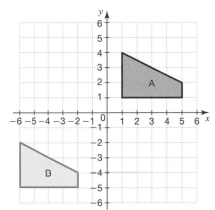

34 Here is triangle *ABC*.

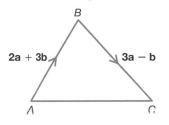

Write down the vector $\overrightarrow{AC}$ in terms of **a** and **b**.

Probability

Basic probability

The probability of an event is how likely it is to occur.

You can describe an event as:

impossible – it will never happen

unlikely – it probably won't happen

likely – it probably will happen

certain – it will definitely happen

In mathematics, the probability of an event is given a numeric value between 0 and 1. The value is given as a decimal, percentage or fraction.

Probabilities can be marked on a probability scale.

For example, the chance of rolling a dice and it landing on an even number is $\frac{1}{2}$, marked with a cross on the scale.

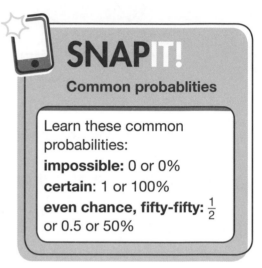

SNAP IT!

Common probablities

Learn these common probabilities:

impossible: 0 or 0%

certain: 1 or 100%

even chance, fifty-fifty: $\frac{1}{2}$ or 0.5 or 50%

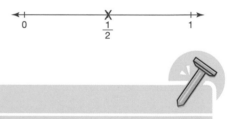

DO IT!

Write down examples of events that have probabilities of 0, 1 and $\frac{1}{2}$.

NAIL IT!

A probability scale is always between 0 and 1. When marking events on a scale, make sure you divide the line equally – so if you want to mark $\frac{1}{3}$ on the scale, measure the line and divide it equally into three parts.

The **relative frequency** of an event can be used as an estimate for the probability of an event when you cannot calculate it another way.

For example, if you wanted to estimate the probability of a piece of toast landing butter side down you could carry out 100 trials. If it landed butter side down on 60 of those trials, the relative frequency of the event would be:

Relative frequency $= \frac{60}{100} = \frac{3}{5}$

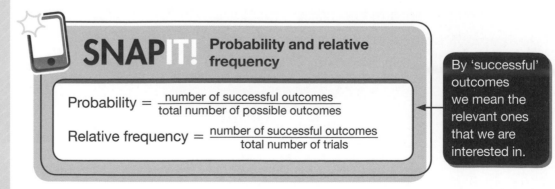

SNAP IT! **Probability and relative frequency**

$\text{Probability} = \frac{\text{number of successful outcomes}}{\text{total number of possible outcomes}}$

$\text{Relative frequency} = \frac{\text{number of successful outcomes}}{\text{total number of trials}}$

By 'successful' outcomes we mean the relevant ones that we are interested in.

WORKIT!

A bag contains 8 red balls, 1 green ball and 11 blue balls.
Timothy takes one ball from the bag without looking.

a Write down the probability that he takes a green ball. Give your answer as a fraction.

> There is 1 successful outcome since there is 1 green ball.
> There are 20 possible outcomes since there are 20 balls in the bag.

$$P(green) = \frac{1}{20}$$

b Timothy says there is an 11% chance he takes a blue ball. Is he correct? Explain your answer.

> If you have to explain your answer you should give a mathematical reason or calculation with your answer.

No, he is not correct - the number of blue balls in the bag is more than half the total, therefore the probability will be greater than $\frac{1}{2}$.

The actual probability of choosing a blue ball is:

$$P(blue) = \frac{11}{20}$$

> Convert this to a percentage.

$$\frac{11}{20} = \frac{55}{100} = 55\%$$

Mutually exclusive events

Two events are mutually exclusive if they cannot happen at the same time. For example, rolling a 1 and rolling a 6 on a dice.

If you list all the possible outcomes of an event and they are all mutually exclusive, then the sum of their probabilities is 1.

Look at the spinner. The possible outcomes are red, blue and yellow, which are all mutually exclusive. You can display the outcomes and their probabilities in a table.

Outcomes	Red	Yellow	Blue
Probability	$\frac{1}{6}$	$\frac{1}{2}$	$\frac{1}{3}$

The sum of the probabilities $= \frac{1}{6} + \frac{1}{2} + \frac{1}{3}$

> If they all add up to 1, you know you have covered all possibilities.

> Convert the fractions so that they all have the same denominator.

$$= \frac{1}{6} + \frac{3}{6} + \frac{2}{6}$$

$$= 1$$

WORKIT!

A box contains milk, dark and white chocolates wrapped in foil paper.

The probability of picking a milk chocolate is 0.3.

The probability of picking a dark chocolate is 0.1.

Calculate the probability of choosing a white chocolate.

$$P(white) = 1 - (0.3 + 0.1)$$

> Since the events are mutually exclusive and the only possible outcomes are choosing a milk, dark or white chocolate, the probabilities sum to 1.

$$= 1 - 0.4$$

$$= 0.6$$

Probability of choosing a white chocolate = 0.6

NAILIT!

A shorthand for writing 'the probability of getting a green ball' is P(green) or P(g).

DOIT!

Write down an example of a pair of events that are mutually exclusive.

WORKIT!

An unfair spinner is numbered 1 to 4. The table shows the probability of scoring each.

Work out the numeric probabilities.

Outcome	1	2	3	4
Probability	$3p$	$2p + 0.2$	p	$2p$

Simplify the equation by collecting like terms.

$3p + (2p + 0.2) + p + 2p = 1$ ◀── Since the events are mutually exclusive, the probabilities must sum to 1.

$8p + 0.2 = 1$

$8p = 0.8$

$p = 0.1$ ◀── Solve the equation to find the value of p.

Substitute the value of p into the expressions.

Outcome	1	2	3	4
Probability	0.3	0.4	0.1	0.2

Probability of an event not occurring

If you know the exact probability of an event occurring you can work out the probability of it not occurring, as they both add up to 1. (It's **certain** that one or the other will happen.)

WORKIT!

The probability that it will rain tomorrow is 0.3. What is the probability of it not raining?

Subtract the probability of it raining from 1.

$1 - 0.3 = 0.7$

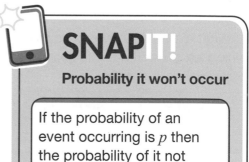

SNAPIT!

Probability it won't occur

If the probability of an event occurring is p then the probability of it not occurring is $1 - p$.

Independent and dependent events

If you roll a dice, the outcome you get on the second roll is unaffected by the outcome on the first roll – the probability of scoring each number remains the same.

However, if you are choosing chocolates from a box and eating them, the first choice you make will affect the second choice. You have changed the number of chocolates in the box, so the probability of each outcome changes.

SNAPIT!

Independence and dependence

Events are independent if the outcome of one event does not affect the outcome of the other.

Dependent events are events whose outcomes do affect one another.

WORKIT!

There are 5 green marbles and 2 red marbles in a bag.

a Jenny chooses a marble from the bag. What is the probability that it is red?

$\frac{2}{7}$ ← There are 2 successful outcomes (red marbles) out of a possible 7 outcomes (total marbles).

b Jenny keeps the first marble, then chooses a second marble. Is the probability that it is red still the same? Explain your answer.

No - since she has removed one of the marbles. There are now only 6 marbles in the bag. Now either 1 or 2 red marbles remain in the bag depending on what the colour of the first marble was. The probability will be $\frac{1}{6}$ if the first ball was red, or $\frac{2}{6}$ if the first ball was green.

DOIT!

Write the following terms on a large piece of paper:

• Probability
• Relative frequency
• Mutually exclusive events
• Independent events
• Dependent events

Next to each term, write down all the things you know about it.

CHECKIT!

1 A fair 10-sided dice numbered 1–10 is rolled.

On the number line, mark with an X the probability of rolling a prime number.

2 A packet of sweets contains 12 toffees, 3 boiled sweets and 10 mints. Eileen takes a sweet from the packet at random. What is the probability that she

a takes a boiled sweet

b doesn't take a toffee?

3 Which of the following pairs of events are mutually exclusive?

a Flipping a coin and getting a head
Flipping a coin and getting a tail

b Rolling a dice and getting a prime number
Rolling a dice and getting an odd number

4 A bag of sweets contains red and green sweets.

Which of these pairs of events are independent?

a Picking one sweet at random, eating it, then choosing another

b Picking one sweet at random, replacing it in the bag, then choosing another

5 A 6-sided dice is biased. The probabilities of rolling the numbers 1 to 5 are shown in the table.

Outcome	1	2	3	4	5	6
Probability	0.1	0.15	0.1	0.02	0.2	

Work out the probability of rolling a 6.

6 A spinner has red, green and blue sections. The probability of it landing on green is twice the probability of it landing on red. The probability of landing on blue is 0.4.

What is the probability of landing on green?

Two-way tables and sample space diagrams

Two-way tables are used when data falls into two different categories.

This table shows the distances (in miles) between three towns in Hampshire:

Petersfield		
12	Waterlooville	
18	9	Portsmouth

The 9 tells you that the distance between Waterlooville and Portsmouth is 9 miles.

This table shows the number of medals won by British competitors in athletics events in an Olympic Games.

	Bronze	Silver	Gold	Total
Men	1	0	3	4
Women	0	1	1	2
Total	1	1	4	6

The 3 in the Gold column tells you that three gold medals were won by the men.
The Total column gives the number of medals won by men and women.
The Total row gives the total number of bronze, silver and gold medals.

WORKIT!

A shop sells three sizes of T-shirts in four different colours. Each T-shirt can be yellow, red, green or blue and is available in three sizes – small, medium or large.

The two-way table shows some information about the number of T-shirts sold in a single week.

1 Find the sum for the Large row.

	Yellow	Red	Green	Blue	Total
Small	3	1		5	12
Medium			14		34
Large	1	7	2	2	
Total		15		10	

2 The first row sums to 12, so 3 green small T-shirts must have been sold. You can now find the total for the Green column.

3 The Red column sums to 15, so 7 medium red T-shirts must have been sold.

4 The Blue column sums to 10, so 3 medium blue T-shirts must have been sold.

5 The Medium row sums to 34, so 10 yellow medium T-shirts must have been sold. You can now find the sum of the Yellow column.

6 The sum of the column totals should be the same as the sum of the row totals: 58. This is a good way to check that you have completed the table correctly.

a How may red T-shirts did the shop sell?

   15 ← Look at the total at the bottom of the 'red' column.

b How many medium T-shirts did the shop sell?

   34 ← Look at the total at the end of the 'medium' row.

c Complete the two-way table.

	Yellow	Red	Green	Blue	Total
Small	3	1	3	5	12
Medium	10	7	14	3	34
Large	1	7	2	2	12
Total	14	15	19	10	58

Sample space diagrams

A sample space (or possibility space) diagram is a type of two-way table that lists all possible outcomes of two events. It can be used to help calculate the theoretical probabilities of outcomes.

NAILIT!

Check if the question uses the word '**and**' or the word '**or**'.
'**And**' means both events must occur.
'**Or**' means one or the other, or both, must occur.

WORKIT!

A fair dice is rolled and a coin is flipped.

a Draw a sample space diagram to show all the possible outcomes.

		Dice					
		1	2	3	4	5	6
Coin	Heads	1, H	2, H	3, H	4, H	5, H	6, H
	Tails	1, T	2, T	3, T	4, T	5, T	6, T

> The cells show the possible outcomes from rolling the dice and flipping the coin.

b Calculate the probability of getting a tail **and** a prime number.

P(tail and prime) $= \frac{3}{12}$

> There are three possible outcomes: 2T, 3T and 5T.

$= \frac{1}{4}$

> There are 12 equally likely outcomes in the table.

c Calculate the probability of getting a head on the coin **or** an even number on the dice.

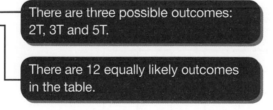

		Dice					
		1	2	3	4	5	6
Coin	Heads	(1, H)	(2, H)	(3, H)	(4, H)	(5, H)	(6, H)
	Tails	1, T	(2, T)	3, T	(4, T)	5, T	(6, T)

> Ring the outcomes with a head or an even number or both.

P(head or even) $= \frac{9}{12}$

> There are 9 equally likely outcomes – 6 with a head and 3 with a tail and an even number.

$= \frac{3}{4}$

185

CHECK IT!

1 Thirty-eight airline passengers are asked for their lunch choices. They may choose chicken, beef or vegetarian option for the main course and fruit or cake for dessert.

Some of the information is shown in the table below.

	Chicken	Beef	Vegetarian
Fruit	12		4
Cake		3	
Total	17	9	

a How many passengers choose the vegetarian option?

b Complete the two-way table.

2 Two fair 6-sided dice numbered 1 to 6 are rolled. The numbers showing on each dice are added together.

A sample space diagram for the possible outcomes is partially drawn below.

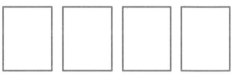

		Dice 1					
		1	2	3	4	5	6
Dice 2	1	2	3	4	5	6	7
	2	3	4	5			
	3						
	4						
	5						
	6						

a Complete the sample space diagram.

b Work out the probability that the score is

   i 3

   ii larger than 10

   iii 1

3 A game is played where four cards are placed face down on a table and a player turns over two at random. Each card has the number 1, 2 or 3 on it. The score is found by adding together the two numbers.

The probability of scoring 6 is $\frac{1}{6}$.

The probability of scoring 2 is $\frac{1}{6}$.

What numbers are on the cards?

Sets and Venn diagrams

In mathematics, a **set** is a collection of things.

For example, the set of prime numbers smaller than 20 is:

$\{2, 3, 5, 7, 11, 13, 17, 19\}$

The set of even numbers is written like this:

$\{2, 4, 6, 8, 10, 12, ...\}$

> These dots tell us the set continues beyond 12.

> The objects in a set are written in curly brackets.

A set of numbers can be described using inequalities. For example:

$A = \{x: 1 < x < 12\}$ is the set of all x values such that x is between 1 and 12.

This could include the numbers 3.21, 8, and 11. It could not include the numbers 1 or 12, since the inequality sign used is $<$ which means 'less than'.

The terms in a set are called **elements**. Sets can be represented by **Venn diagrams** (see Snap it! below).

STRETCH IT!

$M = \{x: -2 < x < 3\}$

and

$N = \{x: 2 \le x < 4\}$

Which whole numbers are in both sets M and N?

> A Venn diagram shows each set as a circle containing the elements in that set. The circles can overlap to show that some elements are in more than one set.

SNAP IT! Set notation

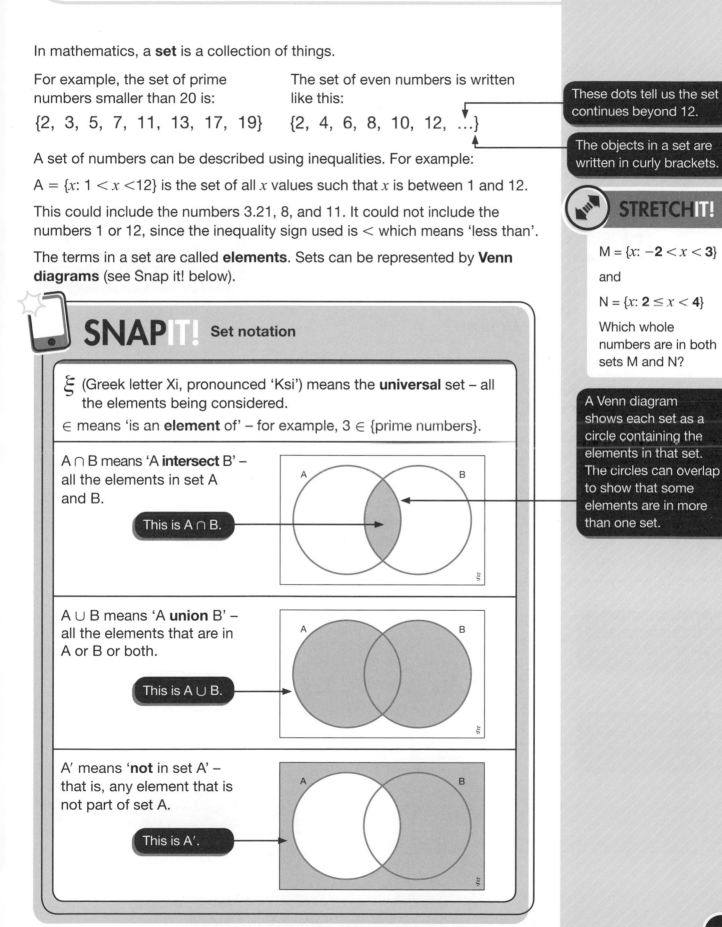

ξ (Greek letter Xi, pronounced 'Ksi') means the **universal** set – all the elements being considered.

$\in$ means 'is an **element** of' – for example, $3 \in \{$prime numbers$\}$.

$A \cap B$ means 'A **intersect** B' – all the elements in set A and B.

> This is $A \cap B$.

$A \cup B$ means 'A **union** B' – all the elements that are in A or B or both.

> This is $A \cup B$.

A' means '**not** in set A' – that is, any element that is not part of set A.

> This is A'.

187

Venn diagrams

A Venn diagram is useful because it shows how the elements of more than one set relate to each other.

Venn diagrams can either list all the elements, or give the number of elements in each set.

> Elements in both sets are written in the overlapping area.

Example 1

This Venn diagram shows:

ξ = {whole numbers smaller than 20}

> This symbol means the universal set – all the elements being considered.

> These sets are numbers smaller than 20, so 20 itself is not included.

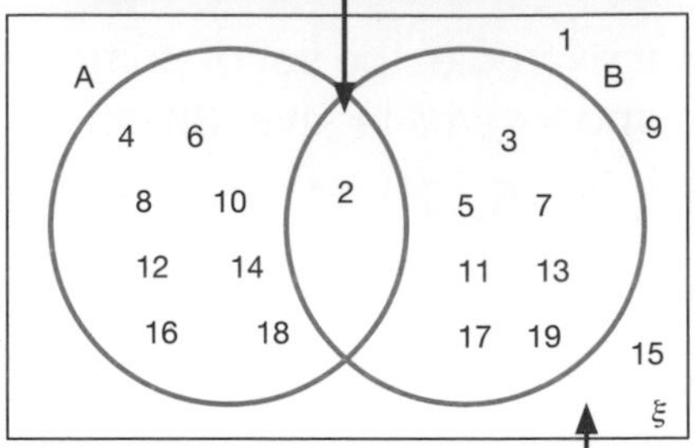

Set A = {even numbers smaller than 20}

Set B = {prime numbers smaller than 20}

In this example, all the elements are listed.

> Elements that do not belong in sets A or B are written outside the circles but are still included since they are in the universal set.

DO IT!

In example 1, how many elements are there in

a A ∪ B

b A ∩ B?

WORKIT!

A group of students is asked whether they play rugby or hockey. R is the set of students who play rugby and H is the set of students who play hockey.

The Venn diagram shows the number of students in each set.

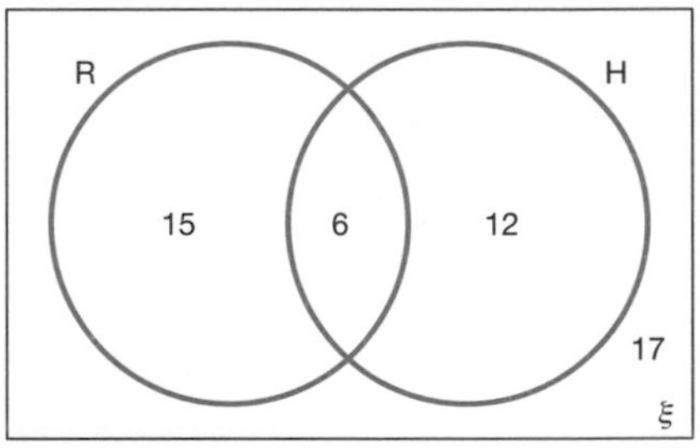

a How many students play neither rugby nor hockey?

17 ← This is the number of students not inside the circles that represent the sets.

b How many students are there in these sets?

i R ∩ H

> R ∩ H represents all the students who play rugby and hockey.

6

ii R ∪ H

15 + 6 + 12 = 33 ← R ∪ H represents all the students who play rugby or hockey or both.

iii R′

> R′ represents all the students who don't play rugby.

12 + 17 = 29

iv ξ

ξ = 15 + 6 + 12 + 17 = 50 ← This means the universal set – all the students in the survey.

Example 2

This Venn diagram shows the number of people studying French, Spanish and German at A level in Year 12 at a school.

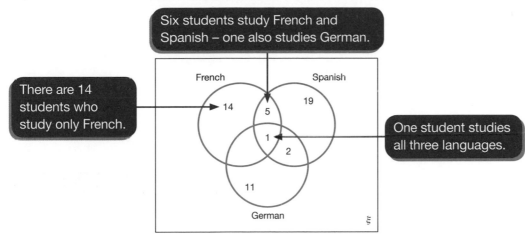

Six students study French and Spanish – one also studies German.

There are 14 students who study only French.

One student studies all three languages.

Calculating probability from Venn diagrams

Venn diagrams can be used to help you work out the numbers you need to use in a probability calculation.

WORKIT!

The Venn diagram shows the number of students in a class with blue eyes and blond hair.

E = {number of students with blue eyes}

H = {number of students with blond hair}

ξ = {students in the class}

A student is selected at random.
Work out the probability that the student

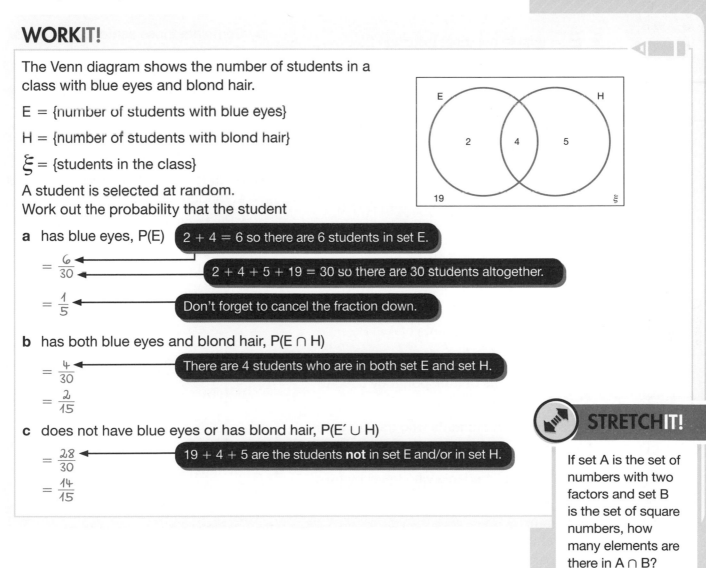

a has blue eyes, P(E)   2 + 4 = 6 so there are 6 students in set E.

$= \frac{6}{30}$ 2 + 4 + 5 + 19 = 30 so there are 30 students altogether.

$= \frac{1}{5}$ Don't forget to cancel the fraction down.

b has both blue eyes and blond hair, P(E ∩ H)

$= \frac{4}{30}$ There are 4 students who are in both set E and set H.

$= \frac{2}{15}$

c does not have blue eyes or has blond hair, P(E′ ∪ H)

$= \frac{28}{30}$ 19 + 4 + 5 are the students **not** in set E and/or in set H.

$= \frac{14}{15}$

STRETCH IT!

If set A is the set of numbers with two factors and set B is the set of square numbers, how many elements are there in A ∩ B?

The addition law

NAIL IT!

In the example shown right, to identify the elements in (E′ ∪ H) you need to consider all elements that are not in E, and all those in H (see shading).

To work out the probability of the events A **or** B, written P(A ∪ B), you cannot simply add together the probability of A and the probability of B.

Thinking about the Venn diagram shown on the right:

$P(E) = \frac{6}{30}$ and $P(H) = \frac{9}{30}$

but $P(E \cup H) = \frac{11}{30}$ which is **not** equal to P(E) + P(H).

This is because if you worked out P(E) + P(H) you would be including the 4 people in the overlapping area twice.

So to work out P(A ∪ B) you can use the formula:

$$P(A \cup B) = P(A) + P(B) - P(A \cap B)$$

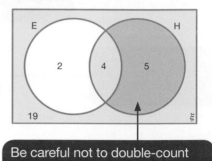

Be careful not to double-count the 5 elements that are just in H.

CHECKIT!

1 ξ = {whole numbers less than 20}

Set A = {multiples of 3 less than 20}

Set B = {multiples of 2 less than 20}

A Venn diagram showing this information has been partially drawn:

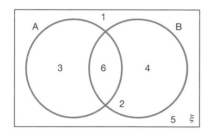

a Complete the Venn diagram.

b Describe the elements of A ∩ B.

2 A year group of 100 students were asked which methods of transport they used to travel to college.

C is the set of students who travel by car.

B is the set of students who travel by bus.

T is the set of students who travel by train.

The Venn diagram shows the number of students in each set.

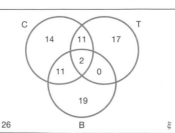

a Complete these sentences. The first is done for you.

C ∪ T is the set of students who travel by car or train or both.

C ∩ T is the set of students who travel by...

C′ ∩ B is the set of students who...

b A student from the group is selected at random. Work out

i P(C)   **ii** P(B ∪ T)   **iii** P(B′ ∩ T)

3 There are 20 elements in sets A and B. Given that $P(A) = \frac{3}{10}$, $P(A \cap B) = \frac{3}{20}$ copy the Venn diagram and mark how many elements are in each region.

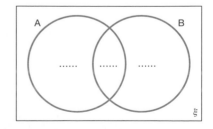

4 Write down the integers that are in the sets.

a {x: 2 < x ≤ 6}

b {x: 0.5 < x < 2.5}

c {x: $\frac{3}{4} \leq x \leq 3\frac{4}{5}$}

Frequency trees and tree diagrams

DOIT!

Using the frequency tree below, work out the percentage of boys that have school lunch.

Frequency trees

A frequency tree shows two or more events occurring and how many times they happen.

This frequency tree shows the number of boys and girls in a school and whether they have packed or school lunches.

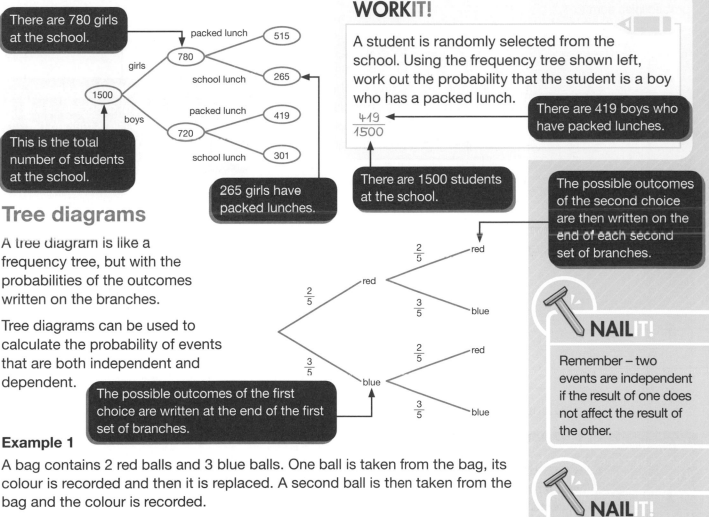

There are 780 girls at the school.

girls 780

packed lunch 515

school lunch 265

1500

This is the total number of students at the school.

boys 720

packed lunch 419

school lunch 301

265 girls have packed lunches.

WORKIT!

A student is randomly selected from the school. Using the frequency tree shown left, work out the probability that the student is a boy who has a packed lunch.

$$\frac{419}{1500}$$

There are 419 boys who have packed lunches.

There are 1500 students at the school.

Tree diagrams

A tree diagram is like a frequency tree, but with the probabilities of the outcomes written on the branches.

Tree diagrams can be used to calculate the probability of events that are both independent and dependent.

The possible outcomes of the first choice are written at the end of the first set of branches.

The possible outcomes of the second choice are then written on the end of each second set of branches.

$\frac{2}{5}$ red

red $\frac{2}{5}$ red

$\frac{3}{5}$ blue

$\frac{3}{5}$ blue

blue $\frac{2}{5}$ red

$\frac{3}{5}$ blue

NAILIT!

Remember – two events are independent if the result of one does not affect the result of the other.

Example 1

A bag contains 2 red balls and 3 blue balls. One ball is taken from the bag, its colour is recorded and then it is replaced. A second ball is then taken from the bag and the colour is recorded.

These events are independent since the outcome of the first selection does not affect the outcome of the second selection.

The tree diagram above shows the probabilities of the possible outcomes.

Each of the different 'routes' through the tree diagram gives a different outcome.

You calculate the probability of each 'route' by multiplying the probabilities together.

red and red $= \frac{2}{5} \times \frac{2}{5} = \frac{4}{25}$ blue and red $= \frac{3}{5} \times \frac{2}{5} = \frac{6}{25}$

red and blue $= \frac{2}{5} \times \frac{3}{5} = \frac{6}{25}$ blue and blue $= \frac{3}{5} \times \frac{3}{5} = \frac{9}{25}$

NAILIT!

If you add together the probability of all of the possible outcomes you should always get 1, since they are mutually exclusive.

$$\frac{4}{25} + \frac{6}{25} + \frac{6}{25} + \frac{9}{25} = 1$$

Example 2

A box of chocolates contains 3 hard centres and 7 soft centres. Maisie chooses a chocolate and eats it; then she chooses a second chocolate.

This tree diagram shows the possible outcomes and the associated probabilities. These events are not independent, since the probability of the second event is affected by the outcome of the first.

> There are only 9 chocolates left in the box when Maisie chooses her second chocolate, so the denominator of the fraction must be 9.

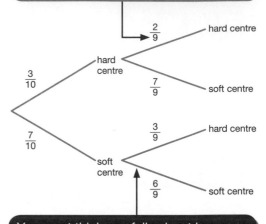

The possible outcomes and probabilities are:

hard and hard $= \frac{3}{10} \times \frac{2}{9} = \frac{6}{90}$

hard and soft $= \frac{3}{10} \times \frac{7}{9} = \frac{21}{90}$

soft and hard $= \frac{7}{10} \times \frac{3}{9} = \frac{21}{90}$

soft and soft $= \frac{7}{10} \times \frac{6}{9} = \frac{42}{90}$

NAILIT!

Check that the probabilities sum to 1.

To find the probability of choosing one of each type of chocolate, add together the probabilities of the outcomes 'hard and soft' and 'soft and hard', since the order in which they are eaten is not specified.

> You must think carefully about how many of each type of chocolate is left in the box when Maisie is making her second choice since she has eaten one chocolate.

$\frac{21}{90} + \frac{21}{90} = \frac{42}{90} = \frac{7}{15}$

WORKIT!

James plays two games at a fair. The probability that he wins the first game is 0.1. The probability that he wins the second game is 0.3.

a Draw a tree diagram to show the possible outcomes.

Game 1	Game 2

0.1 win

0.3 win

0.7 lose

0.9 lose

0.3 win

0.7 lose

> James will definitely either win or lose each game (probability of 1). If the probability that he wins the first game is 0.1 this means the probability he loses this game is 0.9.

b What is the probability that James wins at least one game?

win and win = 0.1 × 0.3 = 0.03
win and lose = 0.1 × 0.7 = 0.07
lose and win = 0.9 × 0.3 = 0.27

0.03 + 0.07 + 0.27 = 0.37

> Work out the probability of each way in which James wins at least once. Remember, this includes him winning both games!

> Add together the probabilities to find the total probability.

CHECK IT!

1 In an orchard there are 347 trees. Of these trees 123 are apple trees and the rest are pear trees.

Of the apple trees 102 are fruiting and of the pear trees 34 are not fruiting.

a Complete the frequency tree to show this information.

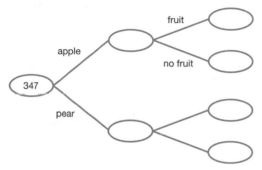

b A tree is chosen at random. What is the probability that the tree is a fruiting pear?

2 A fair 5-sided spinner has the numbers 1 to 5 on it.

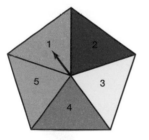

Gary spins the spinner twice and draws a tree diagram showing the possible outcomes.

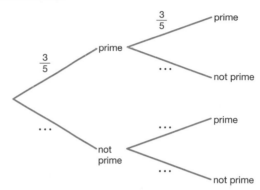

a Complete the diagram.

b Calculate the probability that Gary spins two primes.

c Calculate the probability that Gary spins **at least** one prime.

3 A bag contains 3 glass marbles and 4 plastic marbles. Zac takes one marble from the bag at random, then, without replacing it, he takes a second marble.

a Draw a tree diagram to show the possible outcomes.

b Use the tree diagram to work out the probability that Zac chooses at least one glass marble.

Expected outcomes and experimental probability

Expected outcomes

If you know the theoretical probability of an event occurring then you can calculate how many times it is likely to occur in a given number of trials.

SNAPIT! **Number of occurrences**

Number of times an event will occur = probability of the event × total number of trials

The probability of rolling a 6 on a dice is $\frac{1}{6}$. Therefore, in theory if you roll a dice 12 times you would expect to roll a 6

$\frac{1}{6} \times 12 = 2$ times

In practice this may not actually be the case – this is the **theoretical** probability.

NAILIT!

The more times you repeat an experiment, the closer to the theoretical probability the results will be.

Experimental probability

Sometimes it is necessary to estimate the probability of an outcome using **experimental data**, sometimes called **empirical data**. It is based on observation rather than mathematics or logic.

For example, a student notes the colour of 100 cars passing her house one morning and records the results in a table.

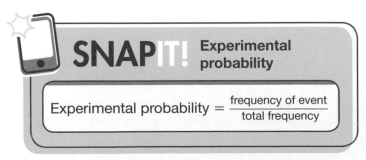

SNAPIT! **Experimental probability**

$$\text{Experimental probability} = \frac{\text{frequency of event}}{\text{total frequency}}$$

Colour	Frequency
Black	23
Blue	12
Red	18
Yellow	1
White	15
Silver	31

Using the data, she **estimates** the probability of the next car being silver as $\frac{31}{100}$.

The experimental probability can be used to estimate how many silver cars you would expect to find in a larger sample.

In a sample of 1000 cars, you would expect $\frac{31}{100} \times 1000 = 310$ cars to be silver.

SNAP IT! Using empirical data

Here are some things you need to remember when using empirical data.

1 The larger the sample, the more accurate the experimental probability.

2 Experimental probability is only an estimate since only a sample has been used.

3 Just because one outcome hasn't occurred doesn't mean it isn't possible.

Bias

If something is biased then one thing is more likely to happen than another – it is 'unfair'.

For example, if a bag contains coloured balls, but the red ones are twice the size of the other colours, then even if you were trying to pick out a ball at random you would be more likely to pick out the red balls. The experiment would be biased.

When conducting probability experiments you should make sure that the experiment is not biased by ensuring every outcome has an equally likely chance.

DO IT!

Make a short list of real-life situations where you have to use experimental probability rather than mathematical probability.

WORK IT!

A biased dice is rolled 20 times and the outcomes are shown in the table below:

Outcome	1	2	3	4	5	6
Frequency	0	2	11	4	1	2

a What is the experimental probability of rolling a 3?

$\frac{11}{20}$ ◄— The number of times a 3 was rolled is 11. The total number of times the dice was rolled is 20.

b Does the fact that a 1 was not rolled mean that it is impossible to roll a 1? Give a reason for your answer.

Not necessarily. Although a 1 was not rolled, this does not mean it is impossible – it just might not have been rolled yet.

c Jenny rolls the dice 100 times. How many times can she expect to roll a 2?

$\frac{2}{20} \times 100 = 10$ ◄— Multiply the experimental probability by the number of times the dice is rolled.

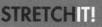 STRETCH IT!

A dice is rolled 12 times and none of the rolls give a 6. Is this enough rolls to say that the dice is biased? Explain your answer.

CHECKIT!

1 The probability of passing your driving test is 0.45. In one month, 300 people sit their driving test. How many people would you expect to pass?

2 A factory produces bags of sweets. Each bag contains 100 sweets. A sample of 10 sweets is chosen from the bag and their colour recorded.

Colour	Frequency
Red	2
Green	6
Yellow	1
Orange	1

Find an estimate for the number of sweets in the bag that are red.

3 A dice is rolled 100 times. If the dice is fair, how many prime numbers would you expect to roll?

4 Three friends carry out an experiment to discover if buttered toast is more likely to land butter side down. Their results are as follows:

	Charlie	Nicolai	James
Butter side down	112	10	28
Butter side up	74	7	19

a Whose results will give the best estimate for the probability of the toast landing butter side down? Justify your answer.

b Charlie carries out the experiment a further 10 times. Use the combined results of the experiment to work out how many times he can expect the toast to land butter side down.

Probability

You may **not** use a calculator for these questions.

1 Approximately 12% of people are left-handed. In a school of 250 students how many would you expect to be left-handed?

2 A fair dice is rolled. Mark the probability of each of these events on a probability scale.

0 ———————————————— 1

A – rolling an even number

B – rolling a factor of 6

C – rolling a 7

D – rolling a number smaller than 8.

3 The probability that it will rain tomorrow is 0.3. What is the probability that it won't rain?

4 Which of the following pairs of events are mutually exclusive:

A – rolling a prime number and rolling an even number on a dice

B – rolling a 3 and rolling a factor of 7 on a dice

C – rolling a 1 and rolling an even number on a dice.

5 A fair 5-sided spinner is labelled with the letters A to E.

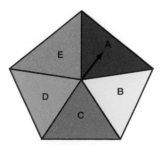

a What is the probability of **not** spinning a vowel (A or E)?

b The spinner is spun 25 times. How many times can you expect to spin a letter B?

6 Complete the two-way table showing lunch choices:

	Pizza	Pasta	Risotto	Total
Cake	12			
Ice cream		11	10	
Total	22	17		50

7 A box of chocolates contains milk, white and dark chocolates.

The table shows the probability of picking each type of chocolate.

Chocolate	Milk	White	Dark
Probability	0.2	$5x + 0.2$	x

Work out the probability of picking a white chocolate.

8 Yan and Ethan have two dice. They think that the dice might be biased.

Yan rolls his dice four times and gets: 2, 2, 6, 1.

a Yan says 'my dice must be biased'. Is he correct? Justify your answer.

b Ethan rolls his dice 50 times and records his results.

Score	1	2	3	4	5	6
Frequency	12	9	16	7	6	0

Ethan decides to roll his dice another 100 times.

Using the results in the table, estimate the number of times he should expect to roll a 2.

197

9 During Holly's journey to work she cycles through two sets of traffic lights. The probability of a red light at the first set of lights is 0.1. The probability of a red light at the second set of lights is 0.5.

The tree diagram shows all the possible outcomes of her journey.

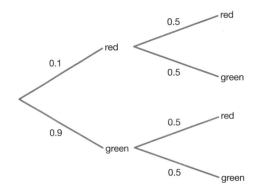

Work out the probability that at least one of the sets of lights is red.

10 Aidyn flips a coin and rolls a dice. If the coin shows heads he doubles the score on the dice. If it shows tails he adds 2 to the score on the dice.

a Complete the possibility space diagram for each possible score.

		Dice					
		1	2	3	4	5	6
Coin	Head	2	4				
	Tail	3	4				

b Find the probability of

   i a score of 4

   ii scoring less than 4.

11 A = {x: whole numbers between 1 and 5}

B = {x: −3 < x ≤ 3}

List the elements that are in A ∩ B.

12 Here is a Venn diagram:

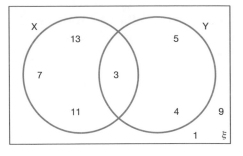

a How many elements are there in

   i X ∪ Y   **ii** X ∩ Y   **iii** Y′?

b A number from the set is chosen at random.

Work out the probability that the number is a member of X.

13 At a stage school all the pupils have piano or singing lessons or both. The Venn diagram shows the number of students who have piano (P) and singing (S) lessons in one class.

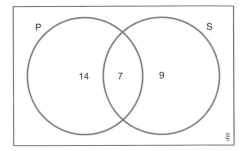

a A student is selected at random from the class.

Work out

   i P(S)

   ii P(S ∩ P)

   iii P(P).

b Yvonne says 'to work out P(S ∪ P), add together P(S) and P(P)'. Is she correct? Explain your answer.

14 A bag contains 2 red balls and 7 green balls.

Alice takes a ball from the bag at random and keeps it, then takes another ball from the bag.

a The tree diagram shows the outcomes and probabilities. Complete the tree diagram.

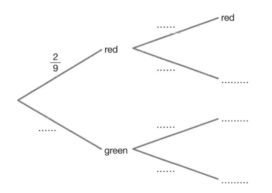

b Use the tree diagram to work out the probability that Alice takes 2 green balls from the bag.

15 A computer randomly generates fractions using the digits 1 to 4 – for example, $\frac{3}{4}$. The numerator is always smaller than the denominator.

a It generates 30 fractions. How many of the fractions can you expect to be less than $\frac{1}{2}$?

b Give a reason why the number of fractions generated that are less than $\frac{1}{2}$ may not be exactly the same as your answer to part a.

16 In a primary school of 300 students 45% are boys.

$\frac{4}{5}$ of the girls and $\frac{2}{3}$ of the boys play sports after school.

A child is picked at random. What is the probability the child plays sport after school?

17 A fairground game has 20 floating ducks on a pond. Players hook out a duck and if the duck has a cross on the bottom it is a winner. There are five ducks with crosses on the bottom.

The game costs £1 to play. If you hook a duck with a cross you win £2.

If 100 people play the game, how much profit should the fairground game make?

You **may** use a calculator for this question.

18 Two students are carrying out a survey to try to work out what percentage of the population is left-handed. Their results are shown in the table below.

	Erica		Milo	
	Male	Female	Male	Female
Right-handed	23	18	51	60
Left-handed	5	4	7	7

a Whose results will give the better estimate for the probability of someone being left-handed? Justify your answer.

b Use the combined results to work out the number of left-handed students you would expect to find in a school with 2000 students.

Statistics

Data and sampling

Statistical investigations involve considering the following questions before collecting any data:

- What problem needs solving?
- What information is needed?
- How should information be collected?
- How should information be presented and displayed?
- How should the information be interpreted?
- What conclusions can be drawn?

When carrying out statistical research about a population you may need to take a **sample**, as it is cheaper and quicker to take a sample than it is to ask the whole population (see Snap it!).

To ensure the sample represents everyone in the country you would need to ensure that you include people:

- of different ages
- of different social and economic backgrounds
- who live in different areas
- of different cultural backgrounds.

You must also make sure the sample is large enough to be a good representation of the population, but not so large that it is too costly or time consuming.

SNAPIT! Data terms

Quantitative Numerical values (for example, age).

Qualitative Describing a quality of something (for example, colour).

Primary data Data collected by the person using it.

Secondary data Data collected by someone else.

Population The complete set of people or things you are interested in.

Sample A small part of the population.

Random sample A sample where every member of the population has an equal chance of being chosen.

Biased sample A sample where a certain type of person or thing is more likely to be chosen.

NAILIT!

If you wished to find out what the most popular chocolate bar in the country is, then the population would be everyone. It would take too long and cost too much to survey everyone in the country so you would take a sample.

STRETCHIT!

If you wanted to find out what the people in your school enjoyed doing in their free time, how would you choose your sample?

WORKIT!

A scientist wishes to investigate the life expectancy of emperor penguins. The population of emperor penguins is approximately 600 000. The scientist takes a sample of 600 birds, all aged between 2 and 3 years old.

Give two reasons why this is not a good sample.

The sample size is small.

The scientist is only sampling $\frac{600}{600\,000} \times 100 = 0.1\%$ of the population.

The sample is biased since the scientist is only considering birds who are between 2 and 3 years old, therefore every member of the population does not have an equal chance of being chosen.

Stratified sampling

If the population you are considering is split into distinct groups, one way of sampling is to ensure the number of people from each group that you sample is proportional to the size of the group. This is called **stratified sampling**.

For example, if there are twice as many German people as there are French people at a meeting and you took a stratified sample, you should sample twice as many German people as French people.

WORKIT!

In a tennis club there are 3500 members, of which 2625 are male.

A stratified sample of 200 members is taken. How many women should be included in the sample?

> First work out how many of the members are female.

$3500 - 2625 = 875$ female members

$\frac{875}{3500} \times 200 = 50$

> Then work out how many women should be in the sample.

50 women should be included in the sample.

Using sampling to describe a population

Once you have collected data using a sample, it can be used to predict the characteristics of the whole population.

For example, a mobile phone company surveys 6000 mobile phone users and finds that 1500 are using their network.

It uses this information to predict that $\frac{1500}{6000} \times 100 = 25\%$ of the population of mobile phone users are using their network.

Since there are approximately 42.4 million mobile phone users in the UK, the company predicts that 25% of 42.4 million = 10.6 million people use their network.

Questionnaires

Having chosen your sample you may wish to design a questionnaire in order to discover the information you need.

When designing questionnaires, follow these rules.

1. Use simple language that is easy to understand.

2. If possible use tick boxes.

3. If you group the data make sure the groups do not overlap. Make them as even as possible and ensure they include all possibilities.

4. Never ask leading questions. For example, the question: 'Fox hunting is cruel, don't you agree?' will encourage the person you are interviewing to answer 'yes'. Instead the question should be: 'What do you think about fox hunting?'

5. Don't ask too many questions. People may be unwilling to answer and it will make carrying out your survey more difficult.

NAILIT!

Try to include mathematical values when commenting on sample size.

NAILIT!

To work out how many of each group to include in a stratified sample:

1. Calculate the percentage or fraction of the population that is in that group.

2. Multiply the sample size by the percentage or fraction.

NAILIT!

The larger the sample, the more accurate the predictions that can be made about the population.

NAILIT!

When working with **continuous data** (like time), you will need to use inequality signs for your groups, like in the example.

When working with **discrete data** (like number of books owned) you do not need to use inequality signs, just make sure your groups don't overlap. For example, to record someone's number of books you could use the groups:

☐ 0 – 19
☐ 20 – 39
☐ 40 – 59
☐ 60 +

Choosing the right question

Here is an example question.

How long do you spend exercising each week?

☐ **0 – 1 hours**   ☐ **1 – 7 hours**   ☐ **7 – 7.5 hours**

This format would create some problems for data collection.

Firstly, people who spent more than 7.5 hours exercising could not be recorded. Secondly, very little meaningful data would be collected as most people would tick the middle box. Lastly, people who spent exactly 1 or 7 hours exercising would not know which box to tick.

A better question would be:

How many hours (h) do you spend exercising each week?

☐ $0 \leq h < 2$ ☐ $2 \leq h < 4$ ☐ $4 \leq h < 6$ ☐ $h \geq 6$

Using inequality signs allows you to specify where the person who exercises for exactly two hours should tick. You should also include an open-ended inequality as the final option so that everyone you survey can be recorded.

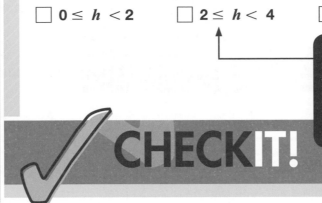

CHECK IT!

1 A class of students wishes to find out about the average temperature in the UK in November. Give one primary and one secondary source of data that they may use.

2 A television company carries out a survey about favourite TV channels. It asks viewers to tell them which channel they watch most often. What type of data is this?

3 Give two reasons why taking a sample is a good way of collecting statistical data.

4 An animal charity carries out a survey on how people feel about wearing real fur. They ask all the people who work for their charity.

  a Explain why this sample would be considered biased.

  b Describe how to choose a sample that is unbiased.

5 In a factory, shampoo is produced in small and large bottles. The table shows the number of bottles produced in a week.

	Large	Small
Number of bottles	1600	400

  a What fraction of the bottles produced are small size?

  b A stratified sample is taken to test quality. If 50 bottles are sampled, how many should be small size?

6 A car manufacturing company makes 800 000 cars a year. A sample of 200 cars is checked on the same date every year to check for faults. The company discovers that 3 of the 200 cars have a fault.

  a How many faulty cars can they expect to produce in a year?

  b Give two reasons why your answer to part a might be unreliable.

7 Zac is carrying out a survey about how long it takes students at his school to travel to school. He asks?

How long do you spend travelling to school in the morning?

0 – 10 minutes

10 – 30 minutes

30 – 60 minutes

  a Give two reasons why this is not a good survey question.

  b Design a better question.

Frequency tables

Grouped data

Frequency tables are a good way of recording data.

This frequency table shows the number of A level subjects taken by a group of students:

Number of A level subjects	Frequency
3	9
4	6
5	5
Total:	20

> You can work out the total number of students by adding together the frequencies:
>
> Total number of students = 9 + 6 + 5 = 20

SNAP IT! Discrete and continuous data

There are two different types of quantitative data:

Discrete data – this can only take particular values – for example, shoe sizes can be 1 or $1\frac{1}{2}$, but not $1\frac{3}{4}$.

Continuous data – this can take any value within a range – for example, your height can be measured to any degree of accuracy, such as 1.4357 m.

To work out the total number of A levels studied, multiply each number of A level subjects by its frequency, then find the sum.

$3 \times 9 = 27$ ◀——— 9 students studied 3 A levels.

$4 \times 6 = 24$ ◀——— 6 students studied 4 A levels.

$5 \times 5 = 25$ ◀——— 5 students studied 5 A levels.

Total A levels studied = 27 + 24 + 25 = 76

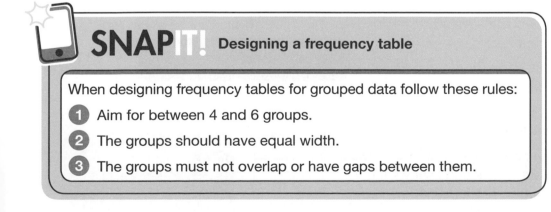

SNAP IT! Designing a frequency table

When designing frequency tables for grouped data follow these rules:

1. Aim for between 4 and 6 groups.
2. The groups should have equal width.
3. The groups must not overlap or have gaps between them.

Discrete data

A railway company wishes to record the number of passengers in each of its train coaches at a particular time of day. The data can only take whole number values (people). This is known as **discrete** data. You can write groups like 1–5, 6–10 and so on.

The frequency table might look like this:

Number of passengers	Frequency
1–25	
26–50	
51–75	
76–100	

The frequency column tells you how many coaches they recorded with a number of passengers in the given range.

Continuous data

A manager of a gym wishes to collect information about the age of customers. The data being collected (age) is continuous so inequalities must be used with no gaps between them.

Age in years (a)	Frequency
$15 \leq a < 30$	
$30 \leq a < 45$	
$45 \leq a < 60$	
$60 \leq a < 75$	

NAILIT!

To determine whether to use $<$ or $\leq$, decide if you **want** that value included in that class – if you do, use $\leq$.

Each class should have one $<$ and one $\leq$.

> $45 \leq a < 60$ means that customer's age is higher than or equal to 45 but lower than 60.

NAILIT!

When grouping data, be careful the groups you are using don't overlap.

For example, if you choose groups of:
0–1 minute
1–2 minutes
for times taken to complete a task, in which group would you record somebody who took exactly 1 minute?

Although grouping data is useful because it enables you to display large amounts of data in a small space, the original values are not recorded and so some of the accuracy of the data is lost.

> First work out how wide the classes will be. Decide whether to divide them into 4, 5 or 6 groups by dividing the total width by 4, 5 and 6.

WORKIT!

A researcher wishes to find out about the average time teenagers spend on their homework each week.

She predicts they will spend between 0 and 25 hours a week on homework.

Design a data collection sheet for her data.

$25 \div 4 = 6.25$
$25 \div 5 = 5$
$25 \div 6 = 4.17$

> Choose a sensible class width that is easy to work with.

Width of class = $25 \div 5 = 5$ hours

Total number of hours (h)	Frequency
$0 \leq h < 5$	
$5 \leq h < 10$	
$10 \leq h < 15$	
$15 \leq h < 20$	
$20 \leq h < 25$	

> Teenagers may do no homework at all, so you must use the $\leq$ sign at the beginning of the class.

CHECKIT!

1 A bus driver records the number of people on her bus at 20 different times during the day. Her results are as follows:

3, 18, 9, 27, 14, 19, 10, 11, 11, 4, 19, 33, 18, 24, 15, 17, 17, 20, 7, 12

Design and complete a grouped frequency table for the data.

2 a A gardener records how many courgettes each of his 15 plants produced. None of his plants produced more than 5 courgettes.

Use this information to complete the frequency table.

Number of courgettes	Frequency
0	1
1	0
2	1
3	1
4	9
5	
6	

b How many courgettes did he grow altogether?

3 A football coach records the hours each of his team spends training each week. He draws up the following table to record the information.

a Give two mistakes he has made.

b The coach says, 'the longest time anyone in the team spends training is 40 hours'. Explain why he might be wrong.

Total number of hours (h)	Frequency
$12 \leq h < 15$	
$16 \leq h < 19$	
$20 \leq h < 30$	
$30 \leq h \leq 40$	

Bar charts and pictograms

Bar charts

A bar chart can be used to represent discrete data.

For example, the data about the number of A levels studied by a group of students can be displayed in a bar chart.

To work out the total number of students, add together the heights of the bars.

Total number of students
$= 9 + 4 + 1 = 14$

Bar line charts

These are exactly the same as bar charts – but with lines drawn instead of bars.

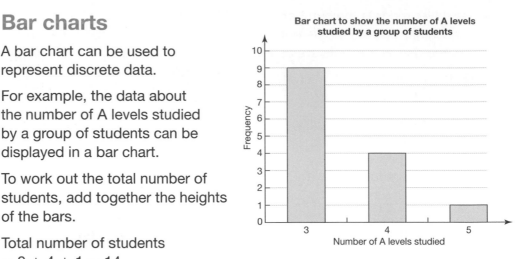

Bar chart to show the number of A levels studied by a group of students

Bar line chart to show the number of A levels studied by a group of students

Comparative bar charts

A comparative bar chart can be used to easily compare sets of data.

From this comparative bar chart, you can see at a glance that there are more boys in both the Maths and Physics classes.

The key is essential.

Comparative bar chart showing the number of boys and girls taking different A level options

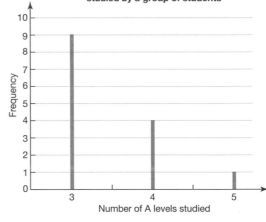

12 girls study Maths.
15 boys study Maths.

DOIT!

From the comparative bar chart on the right, how many more boys study Physics than girls?

Composite bar charts

This composite bar chart displays the same information as the comparative bar chart. It is easier to see at a glance the total number of students in each lesson, but more difficult to identify the exact number of boys or girls in each.

Bar charts for grouped discrete data

A bar chart can also be used to display grouped discrete data.

This bar chart shows the number of students in evening classes offered by a further education college.

The total number of classes can be found by adding together the heights of the bars:

total number of classes
= 9 + 11 + 1 = 21

Pictograms

Pictograms can be used to display discrete data.

For example, here is a pictogram displaying numbers of A levels that students take.

The key states each symbol represents 2 students, so the total number of students in the sample can be calculated by counting the symbols and multiplying by 2.

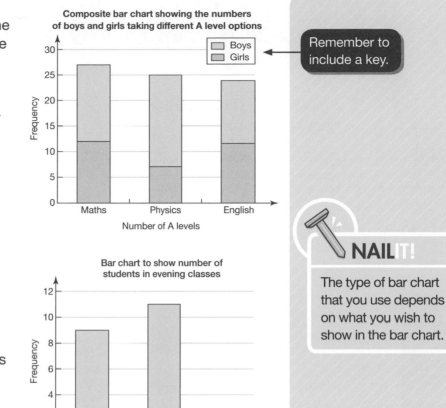

Composite bar chart showing the numbers of boys and girls taking different A level options

Bar chart to show number of students in evening classes

Number of A levels	Number of students
3	⬭⬭⬭⬭⬭
4	⬭⬭
5	⬭

⬭ = 2 students

Remember to include a key.

NAIL IT!

The type of bar chart that you use depends on what you wish to show in the bar chart.

DO IT!

How many people does this pictogram represent altogether?

NAIL IT!

Drawing pictograms
- Use only one symbol.
- All symbols must be the same size.
- Choose a symbol that you can easily halve or quarter if necessary.
- Include a key.
- Space symbols evenly.
- Label the axis.

DO IT!

Without looking at the book, list all the things you should remember when drawing:

1 bar charts
2 pictograms.

CHECK**IT!**

1 A group of students were asked how many pets they owned.

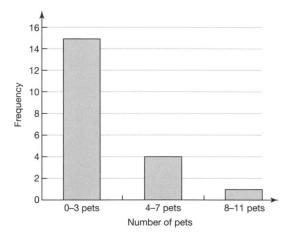

a How many students were surveyed?

b What percentage of students owned four or more pets?

2 A group of college students were asked how many sports they played regularly. The results are shown in the composite bar chart.

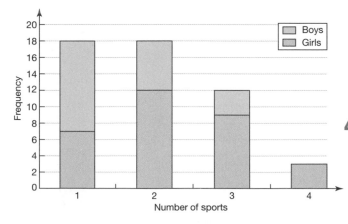

a How many more boys than girls played only one sport?

b What percentage of students surveyed were boys?

c Jasmine says that the proportion of boys to girls who play two sports is larger than the proportion of boys to girls who play three sports. Show that this is true.

3 A shop records the size of shoes it sells in one day. The bar line chart displays some of the information.

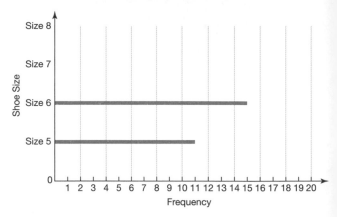

The shop sells a total of 50 shoes on the day. It sells twice as many size 7 shoes as size 8 shoes. Copy and complete the chart.

4 The pictogram shows the colours of bikes manufactured at a factory.

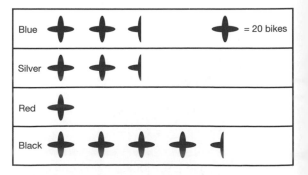

a How many more black bikes than red bikes were made?

b What proportion of the bikes made were silver?

Pie charts

Drawing pie charts

SNAPIT!   Drawing a pie chart

To draw a pie chart, follow these steps:

1 Work out the total number of items your pie chart represents.

2 Divide 360° by the total number of items. ◄

3 Multiply the angle (your answer to Step 2) by the number of items in each group – this gives you the angle for each group.

4 Construct the pie chart – use a protractor to measure the angles.

5 Label the regions or include a key.

> This gives the size of pie slice per item.

WORKIT!

The favourite sport of a group of students is recorded in this frequency table. Calculate the angles of a pie chart to represent the data.

Favourite sport	Frequency
Football	22
Rugby	15
Tennis	12
Swimming	11

Total number of students = 22 + 15 + 12 + 11

= 60 ◄ *Find the sum of the frequencies.*

360° ÷ 60 = 6° ◄ *Divide 360° by the sum of the frequencies. 6° represents one student.*

Football = 22 × 6° = 132°

Rugby = 15 × 6° = 90°

Tennis = 12 × 6° = 72° ◄ *Calculate the angle needed for each sport.*

Swimming = 11 × 6° = 66°

NAILIT!

Check the angles sum to 360° – if they don't, you have made a mistake somewhere!

For this example,
132° + 90°
+ 72° + 66° = 360°

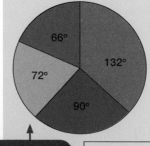

66°
72°
132°
90°

> You would not actually write the angles on your chart.

■	Football
■	Rugby
□	Tennis
■	Swimming

Interpreting pie charts

The pie chart shows the distribution of different flavours of 30 sweets in a packet.

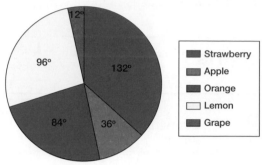

You can work out how many sweets there are of each flavour by working out how many degrees represent each sweet.

$360° ÷ 30 = 12°$ ← Divide 360° by the number of sweets. 12° represents each sweet.

Strawberry $= 132 ÷ 12 = 11$ sweets

Apple $= 36 ÷ 12 = 3$ sweets

Orange $= 84 ÷ 12 = 7$ sweets

Lemon $= 96 ÷ 12 = 8$ sweets

Grape $= 12 ÷ 12 = 1$ sweet

$11 + 3 + 7 + 8 + 1 = 30$ sweets ← Check your answer – the total number of sweets should be 30.

Divide each angle by 12 to work out how many sweets each section represents.

DO IT!

Draw a flow chart to describe how to construct a pie chart.

STRETCH IT!

If the angles are not whole numbers, how would you deal with that?

WORKIT!

The pie chart shows the ages (x) of 120 people who use a swimming pool.

a How many people are between 18 and 40?

20% are in the group

$18 < x ≤ 40$

10% of 120 = 12

20% = 2 × 12 = 24 people.

■	$0 < x ≤ 18$
■	$18 < x ≤ 40$
■	$x > 40$

b How many more people using the pool are under 18 than are between 18 and 40?

70% of 120 = 7 × 12

= 84

$84 − 24 = 60$ more people ← Subtract the number of people between 18 and 40 from the number of people under 18.

1 Students at a school choose to study French, Spanish or German.

Draw a pie chart to represent this data.

Language	Number of students
French	27
Spanish	42
German	21

2 The pie chart shows the annual salaries of 240 students leaving university.

a How many students earned over £40 000 a year?

b What proportion of students earned less than £30 000?

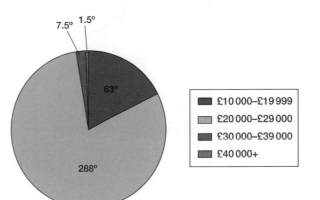

£10 000–£19 999
£20 000–£29 000
£30 000–£39 000
£40 000+

3 Stefan uses a pie chart and a bar chart to display data he has collected about the number of cars owned per household.

Number of Cars	Frequency	Angles
0	6	18
1	21	63
2	65	195
3	18	54
4	10	30

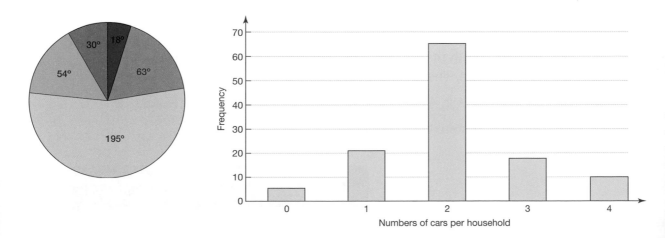

a How many households owned 3 or more cars?

b Which chart is it easier to read this from? Explain why.

Stem and leaf diagrams

A stem and leaf diagram shows data split into a 'stem' and a 'leaf'.

Stem and leaf diagram showing the ages of people attending a yoga class

The 'stem' is the tens digit, the 'leaf' is the units.

The data is ordered in the diagram

The numbers in the 'leaves' should be evenly spaced so that the shape of the data is easily seen.

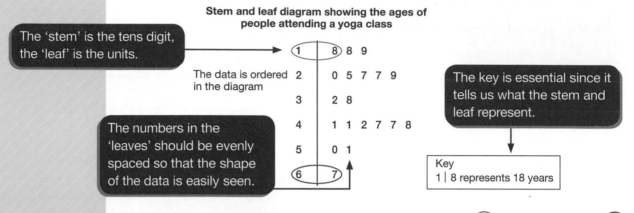

1	8 8 9
2	0 5 7 7 9
3	2 8
4	1 1 2 7 7 8
5	0 1
6	7

The key is essential since it tells us what the stem and leaf represent.

Key
1 | 8 represents 18 years

It is easy to see that the youngest person in the class is 18 and the oldest is 67.

Back-to-back stem and leaf diagrams

A back-to-back stem and leaf diagram shows two sets of data, allowing comparisons to be made.

Stem and leaf diagram showing the exam results of a class of students

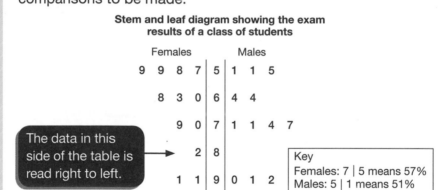

Females		Males
9 9 8 7	5	1 1 5
8 3 0	6	4 4
9 0	7	1 1 4 7
2	8	
1 1	9	0 1 2

The data in this side of the table is read right to left.

Key
Females: 7 | 5 means 57%
Males: 5 | 1 means 51%

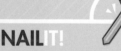

NAILIT!

Check you have included all the data by counting how many pieces of data are in the stem and leaf diagram and comparing against the stated total.

Next draw the stem and leaf diagram, ordering the leaves.

WORKIT!

The length of 15 slugs in centimetres is measured and recorded. The results are as follows:

3.1, 5.6, 4.2, 4.7, 5.0, 2.9, 3.3, 4.1, 5.2, 3.2, 4.1, 3.5, 2.8, 4.5, 5.0

Draw a stem and leaf diagram to display the data.

2	9 8
3	1 3 2 5
4	2 7 1 1 5
5	6 0 2 0

First draw an 'unordered' stem and leaf diagram, getting the correct leaves on the correct stem.

2	8 9
3	1 2 3 5
4	1 1 2 5 7
5	0 0 2 6

Key
2 | 8 represents 2.8 cm

SNAP IT! Stem and leaf diagram

Follow these steps to construct a stem and leaf diagram:

1. Decide on the 'stem' and the 'leaf'.

2. Draw an unordered stem and leaf diagram.

3. Draw the stem and leaf diagram, making sure that numbers are evenly spaced.

4. Include a key.

CHECK IT!

1 The stem and leaf diagram shows the weights of babies born in a maternity unit.

Females		Males
	1	9
6 3 3	2	0 2 2
8 8 5 1	3	5
3 3 0	4	0 1

Key
Females: 3 | 2 means 2.3 kg
Males: 1 | 9 means 1.9 kg

a How many male babies were born?

b What is the difference in weight between the heaviest girl born and the heaviest boy born?

2 The stem and leaf diagram below has been drawn badly. Redraw it correctly, explaining what was incorrect in the original diagram.

Age of people using a dentist

2	1 1 1 0 0 0 0
3	5527
4	22 1 6

3 A bar chart and a stem and leaf diagram are used to display information about the number of passengers using a daily ferry service.

0	3 3 3 4 9
1	2 3 4 4 8 9 9 9
2	2 2 3 5 7

Key
2 | 3 means 23 passengers

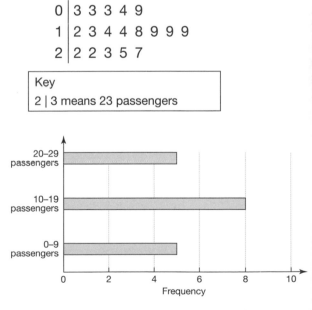

From which diagram can you read:

a the smallest number of passengers

b the trends in the data?

Give a reason for your answers.

Measures of central tendency: mode

'Measures of central tendency' are often referred to as 'averages'. There are several different types of measures of central tendency.

The **mode** is the value that occurs most often in a set of data.

If the data is grouped, the **modal class** is the one with the highest frequency.

The mode is the only measure of central tendency (average) that can be found from both quantitative and qualitative data, it is easy to find and can show what the most 'popular' value or thing is.

Finding the mode from a list of data

The colour of 10 cars is recorded, with results as follows:

blue, green, white, silver, silver, blue, red, silver, white, black

Which is the modal colour?

The mode is silver since this occurs three times – more than any other colour.

Sometimes there is no mode since no data value is repeated more than once or each value is repeated the same number of times.

Sometimes there is more than one mode, but we do not usually give more than two modes, instead we say there is no mode.

The age, in years, of swimmers in a race was recorded:
11, 11, 13, 11, 13, 12, 12, 11, 13, 12, 13, 12, 11, 13

The modal values are 11 and 13. This is not a particularly useful piece of information.

Finding the mode from a frequency table

If the data is displayed in a frequency table, the mode or modal group is the one with the highest frequency.

For example, this table shows the ages of students taking GCSE maths at a school.

Age of students taking GCSE maths	Frequency
15	32
16	144
17	19

The modal age of students taking GCSE maths is 16, since this is the age of the largest number of students.

The second table shows the ages of students at a school.

The modal class is $11 < a \le 12$ since this is the class with the largest number of students.

Age of students (a)	Frequency
$10 < a \le 11$	112
$11 < a \le 12$	125
$12 < a \le 13$	110
$13 < a \le 14$	109
$14 < a \le 15$	100
$15 < a$	92

Finding the mode from a bar chart

It is easy to identify the mode from a bar chart. It is the tallest bar.

Here the mode is maths since this is the subject that the greatest number of students study.

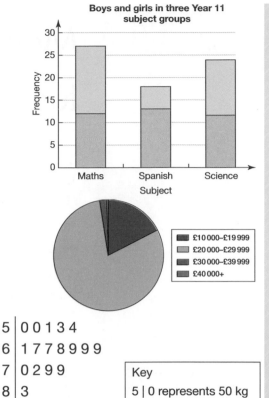

Finding the mode from a pie chart

The modal value is the one represented by the largest sector.

In the survey of students' starting salaries from page 211, the modal class is £20 000–£29 999, since this is the largest segment (see pie chart right).

Finding the mode from a stem and leaf diagram

Look for the value that appears most often – remember the value must have the same stem **and** leaf.

This stem and leaf diagram shows the weight (in kg) of patients at hospital.

The mode is 69 kg.

```
5 | 0 0 1 3 4
6 | 1 7 7 8 9 9 9
7 | 0 2 9 9
8 | 3
```

Key
5 | 0 represents 50 kg

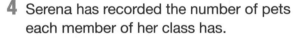

CHECKIT!

1 Five numbers have a mode of 12.2. Given that two of the numbers are 11.9, what can you say about the other numbers?

2 A group of students records the length of time they spend on the internet in a day.

What is the modal class?

Time, t (hours)	Frequency
$0 \leq t < 1$	13
$1 \leq t < 2$	62
$2 \leq t < 3$	12
$3 \leq t < 4$	5
$4 \leq t < 5$	8

3 The back-to-back stem and leaf diagram shows the age of contestants in a singing competition.

What is the modal age of the female competitors?

```
   Male |   | Female
   9 9 8 7 | 1 | 6 6 7 7 7 9
7 7 5 2 2 2 1 0 | 2 | 0 0 1 1 3
       2 1 | 3 | 0
```

Key
Males: 7 | 1 represents 17 years
Females: 1 | 6 represents 16 years

4 Serena has recorded the number of pets each member of her class has.

Number of pets	Frequency
0	0
1	7
2	14
3	8
4	0
5	0

Serena says, 'the modal number of pets is 0 since it appears three times in the frequency column'.

Max says that is wrong.

Who is correct? Explain why.

Measures of central tendency: median

The median is the middle value of an ordered set of data.

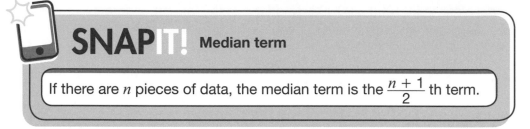

SNAPIT!   **Median term**

If there are n pieces of data, the median term is the $\frac{n+1}{2}$ th term.

WORKIT!

Find the median of these values: 2, 3, 1, 4, 4, 6, 9, 2, 7

1   2   2   3   ④   4   6   7   9

Median term $= \frac{9+1}{2} = 5^{th}$ ← There are nine terms in the list. The median is the 5th term.

The median is 4.

NAILIT!

You must make sure the items of data are ordered before finding the median.

Finding the median from a frequency table

If data is grouped, you must first work out the total number of pieces of data.

The table shows the number of siblings a group of students have.

Number of siblings	Frequency
0	3
1	5
2	7
3	1
Total:	16

If it isn't provided, first find the sum of the frequencies.

Median term $= \frac{16+1}{2} = 8.5$th term

The median term lies between the 8th and 9th terms.

You could list all the terms:

0   0   0   1   1   1   1   1   2   2   2   2   2   2   2   3

The 8th and 9th terms are 1 and 2 – therefore the median is $\frac{1+2}{2} = 1.5$.
You could work out that the 8th and 9th terms are 1 and 2 by considering the frequencies in the table.

NAILIT!

If the median term lies between two values, add together the values and divide by 2.

Finding the median of grouped data

Finding the actual value of the median from grouped data is beyond the scope of this revision guide, but you can find the median class.

WORKIT!

The length of time a group of people spend exercising each week is recorded. Which is the median class?

$7 + 15 + 18 + 3 = 43$ ◄─── Find the sum of the frequencies.

$\text{Median term} = \frac{43 + 1}{2} = 22\text{nd term}$

The 22nd term is in the class $2 \leq t < 4$

Median class $= 2 \leq t < 4$

Time, t (hours)	Frequency
$0 \leq t < 2$	7
$2 \leq t < 4$	15
$4 \leq t < 6$	18
$6 \leq t < 8$	3

Finding the median from a stem and leaf diagram

Work out how many pieces of data there are by counting the leaves on a stem and leaf diagram.

Stem and leaf diagram showing the ages of people attending a yoga class

```
1 | 8 8 9
2 | 0 5 7 7 9
3 | 2 8
4 | 1 1 2 7 7 8 9
5 | 0 1
6 | 7
```

Key
1 | 8 represents 18 years

There are 20 leaves.

Median term $= \frac{20 + 1}{2} = 10.5$th term.

Therefore the median term lies between the 10th and 11th terms.

10th term = 38   11th term = 41

Median $= \frac{38 + 41}{2} = 39.5$

CHECKIT!

1 Find the median of the list of numbers below

3.1   9.2   8.7   6.5   4.3   2.9

2 A group of students records the number of wild birds they see in a week.

What is the median class?

Birds, b	Frequency
$0 \leq b < 2$	29
$2 \leq b < 4$	28
$4 \leq b < 6$	30
$6 \leq b < 8$	3
$8 \leq b < 10$	10

3 The back-to-back stem and leaf diagram shows the marks of two groups of students in a maths test:

```
   Group A        Group B
       8 8 | 7 | 0 0 1 4 5 9
 9 9 8 5 2 0 0 | 8 | 2 2 3 9
           3 | 9 |
```

Key
Group A:
8 | 7 represents
78 marks
Group B:
7 | 0 represents
70 marks

a Find the median score for both groups.

b Use the median to decide which group you think did better in the test.

Measures of central tendency: mean

When we talk about the average we are usually talking about the mean.

SNAP IT! Mean

$$\text{Mean} = \frac{\text{sum of all data values}}{\text{total number of values}}$$

WORKIT!

Calculate the mean of 10, 15, 19, 7 and 4.

$$\text{Mean} = \frac{10 + 15 + 19 + 7 + 4}{5}$$
$$= \frac{55}{5}$$
$$= 11$$

Divide by 5 since there are five data values.

Calculating the mean from a frequency table

Siblings are brothers and sisters.

This table shows the number of siblings that a group of students have.

In order to work out the mean number of siblings you need to calculate:

total number of siblings ÷ total number of students

Number of siblings	Frequency
0	3
1	19
2	13
3	1

Draw a third column on the table:

Number of siblings	Frequency	Number of siblings × frequency	
0	3	0 × 3 = 0	← 3 people had 0 siblings.
1	19	1 × 19 = 19	← 19 people had 1 sibling.
2	13	2 × 13 = 26	
3	1	3 × 1 = 3	
Total:	36	48	← The total number of siblings.

The total number of students.

$$\text{Mean} = \frac{48}{36} = 1\frac{1}{3} \text{ siblings}$$

STRETCH IT!

Is this a sensible answer? You know the answer must lie between 0 and 3 siblings. But if you wanted a whole number you might choose to use the mode instead.

Calculating the mean from a bar chart

This bar chart shows the number of pets owned by students in a class.

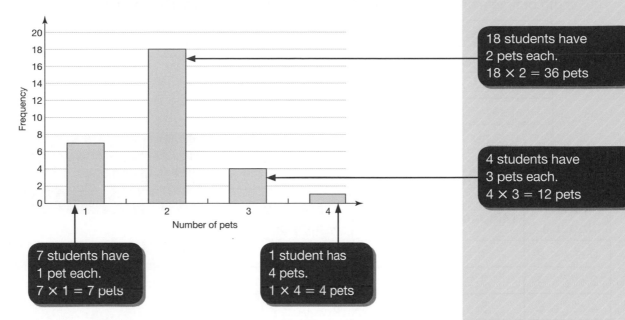

18 students have 2 pets each. $18 \times 2 = 36$ pets

4 students have 3 pets each. $4 \times 3 = 12$ pets

7 students have 1 pet each. $7 \times 1 = 7$ pets

1 student has 4 pets. $1 \times 4 = 4$ pets

To work out the mean number of pets, work out:

total number of pets ÷ total number of students

The total number of students is the sum of the heights of the bars:

$7 + 18 + 4 + 1 = 30$

To work out the total number of pets, multiply the height of each bar by the number of pets it represents:

Total number of pets $= (1 \times 7) + (2 \times 18) + (4 \times 3) + (1 \times 4)$

$$= 7 + 36 + 12 + 4 = 59$$

Mean number of pets $= \dfrac{59}{30}$

$$= 2.0 \text{ pets (to 1 decimal place)}$$

STRETCH IT!

Which measure of central tendency would you use in each of the following situations?

a Average shoe size

b Average length of time spent waiting for a bus

c Average amount of money spent buying a new car

Estimating the mean of grouped data

You cannot find the exact mean from grouped data since you do not know what the values of the original data were. However, you can find an estimate for the mean.

Use the midpoint of the class as the data value.

WORKIT!

The length of time a group of people spend exercising each week is recorded.

Time, t (hours)	Frequency
$0 \leq t < 2$	7
$2 \leq t < 4$	15
$4 \leq t < 6$	18
$6 \leq t < 8$	3

Calculate an estimate for the mean to one decimal place.

> Add another column to calculate the midpoint of the class. Use this as an estimate of the actual time spent exercising.

Time, t (hours)	Midpoint	Frequency	Frequency × midpoint
$0 \leq t < 2$	1	7	$1 \times 7 = 7$
$2 \leq t < 4$	3	15	$3 \times 15 = 45$
$4 \leq t < 6$	5	18	$5 \times 18 = 90$
$6 \leq t < 8$	7	3	$7 \times 3 = 21$
	Total:	43	163

> 7 people spend 1 hour exercising.

> The total time spent exercising.

$$\text{Mean} = \frac{163}{43} = 3.8 \text{ hours}$$

> The total number of people surveyed.

CHECKIT!

1 Andrea records the length of time she spends waiting for her bus in one month.

Time, t (mins)	Frequency
$0 \leq t < 4$	12
$4 \leq t < 8$	3
$8 \leq t < 12$	5

  a Calculate an estimate for the mean length of time Andrea waits each day.

  b Explain why your answer is an estimate.

2 In a quiz the mean score of five students is 9. Another student scores 6 marks.

What is the new mean?

3 The mean of three numbers is 5. One of the numbers is 5.

Yasmina says that the other numbers must also be 5.

Is she correct? Explain your answer.

Range

The range is a measure of how spread out data is.

Finding the range from a bar chart

Here's a bar chart showing the number of towns students have lived in.

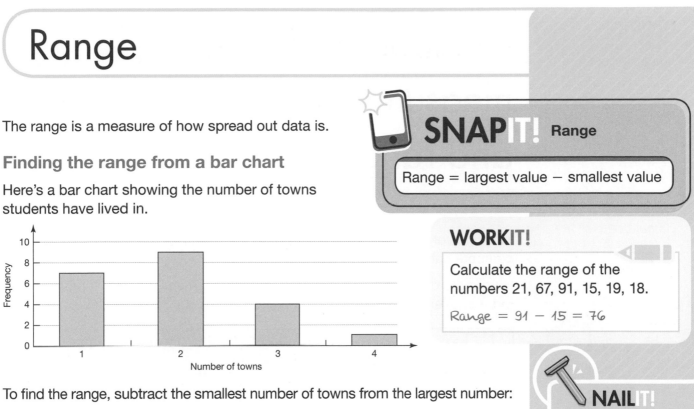

To find the range, subtract the smallest number of towns from the largest number:

Range = 4 − 1 = 3

Finding the range from a stem and leaf diagram

SNAPIT! Range

Range = largest value − smallest value

WORKIT!

Calculate the range of the numbers 21, 67, 91, 15, 19, 18.

Range = 91 − 15 = 76

NAILIT!

Don't subtract the frequencies – these are **not** the data values.

WORKIT!

This stem and leaf diagram shows the exam results of a group of students. Find the range of marks for the female students.

Range = highest mark − lowest mark

= 91 − 57 = 34

Females		Males
9 9 8 7	5	1 1 5
8 3 0	6	4 4
9 0	7	1 1 4 7
2	8	
1 1	9	0 1 2

Key
Females: 7 | 5 means 57%
Males: 5 | 1 means 51%

CHECKIT!

1 Calculate the range of the following values 3.1, 4.2, 9.5, 8.7, 5.2, 9.1, 9.2, 4.9, 0.7, 1.4

2 The bar chart shows the age of students studying A level German at a school.

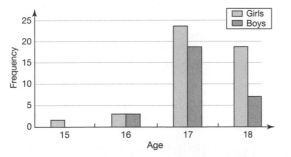

Work out the range of ages of the
a girls   b boys.

3 The times in seconds taken by two athletes to run 100 m are shown.

Athlete A	13.2	14.5	13.0	13.9	14.7	15.2
Athlete B	14.3	14.8	14.5	15.2	14.9	15.0

Which athlete has the greater range of times?

4 In a science test the range of marks is 20. The lowest mark is 45%.

Another student takes the test and the range increases to 30. What are the possible marks the student could have achieved?

Comparing data using measures of central tendency and range

Measures of central tendency and range are useful as they enable you to compare data.

Here is the data on the distances, to the nearest metre, that two sportsmen throw a javelin.

	Mean length of throw	Modal length of throw	Median length of throw	Range
Sportsman A	81 m	82 m	82 m	16 m
Sportsman B	81 m	98 m	79 m	38 m

Which sportsman would you choose for your team?

Either answer can be correct as long as you justify your answer using the data.

You might choose Sportsman A since his range is lower, therefore he is more consistent in the length of his throws.

You might choose Sportsman B since his modal throw is much higher than Sportsman A's.

DOIT!

Make flash cards with 'Mean', 'Median' and 'Mode' on them and write down the advantages and disadvantages of each.

Advantages and disadvantages of each measure of central tendency

	Advantages	Disadvantages
Mode	1 Quick and easy to calculate. 2 The mode of qualitative data can be found. 3 You can find the mode or modal class of discrete and continuous data. 4 Extreme values do not affect the mode.	1 It does not represent the whole data set since extreme values are ignored. 2 There may be more than one mode. 3 There may be no mode – for example, if each value occurs only once.
Median	1 Extreme values do not affect the median.	1 It can be time-consuming to find for large data sets. 2 It does not take into account any extreme values.
Mean	1 All the data is used to find the answer.	1 Extreme values can distort the data. 2 It can be time-consuming to calculate.

Your choice of measure of central tendency can affect the conclusions you draw from a statistical investigation.

WORKIT!

A battery manufacturer wishes to show how long-lasting its batteries are.

A sample of five batteries is tested and the length of time they last for, in the same device, is recorded in hours. The results are as follows:

15.4, 18.9, 2.4, 14.7, 19.6

a Which average would the company use in its advertising? Explain your answer.

$$Mean = \frac{15.4 + 18.9 + 2.4 + 14.7 + 19.6}{5} = 14.2$$

Median:

2.4, 14.7, 15.4, 18.9, 19.6 ← Write the data in order first.

Median = 15.4

Mode: There is no mode

The company would use the median since this is the highest value.

b A rival company uses the range of the data to show how unreliable the batteries are. Explain why they use this calculation.

The range is 19.6 − 2.4 = 17.2

This shows the lifespan of the batteries can vary greatly.

NAILIT!

If asked to explain your answer, use calculations to support your decision if possible.

✓ CHECKIT!

1 The length of time children spend on a computer game over one weekend is recorded:

32 mins, 29 mins, 18 mins,
41 mins, 362 mins, 19 mins

Work out:

a i the mean   ii the median.

b Why are the averages so different?

2 A school uses the mean to calculate the average age of students in the year. Give a reason why they might have chosen to use this measure of central tendency.

3 A car manufacturer records the times it takes two different cars to accelerate from 0 to 60 miles per hour.

They carry out several experiments. The results are shown below.

Car A: Mean = 5.3 seconds
       Range = 0.9 seconds

Car B: Mean = 4.2 seconds
       Range = 2.3 seconds

Which car would you recommend? Justify your answer.

4 A children's shoe shop records the size of wellington boots sold one morning.

Shoe size	Pairs of boots sold
3	2
4	5
5	18
6	2

a Which average would be a useful measure of central tendency?

b Explain why you have chosen this average.

Time series graphs

DO IT!

Use this graph to work out:

a the highest temperature

b the lowest temperature

c the range of temperatures.

A time series graph is used to show how data changes over time and can be used to spot trends over time.

This graph shows how the temperature changes during a single day.

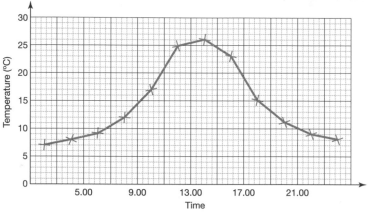

The highest and lowest temperatures can be read easily from the graph.

Take care when reading particular values from a time series graph. Remember that the line joining two points does not show accurate values between those points.

Plotting a time series graph

NAIL IT!

Although a trend can be spotted using a time series graph, you must be careful when predicting how the graph will behave beyond the time plotted. You do not know that this is how the data will behave because it is outside the scope of what that has been recorded.

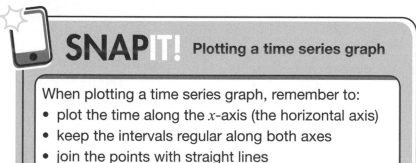

SNAP IT!   Plotting a time series graph

When plotting a time series graph, remember to:
- plot the time along the x-axis (the horizontal axis)
- keep the intervals regular along both axes
- join the points with straight lines
- label the axes.

WORK IT!

The height of a tomato plant is measured over 20 days.

a What height were the seedlings on day 11?

6 cm ◀— Draw a vertical line from the x-axis at day 11, and a horizontal line from the point the vertical line touches the time series line, and read off the value.

b What height would you expect the seedling to be on day 21? Explain whether you think your answer is accurate.

16 cm. The prediction is unreliable since it is outside the scope of the data values recorded.

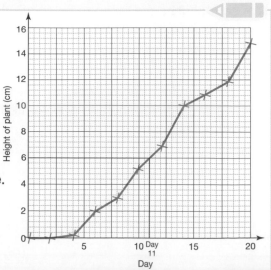

Constructing a table for a time series

To construct a table to display a time series, make sure that:

- the time intervals are regular
- all the relevant times are included.

For example, if you want to record air temperature over a 24-hour period, you might produce a table like this one.

Time	0	2	4	6	8	10	12	14	16	18	20	22	24
Temperature													

> The 24-hour period has been split into 12 equal intervals.
>
> You could split it into more or fewer intervals as long as they are the same length.

CHECKIT!

1 The table shows the speed of an accelerating car over a 1-minute interval.

Draw a time series graph to represent the data.

Time (s)	0	10	20	30	40	50	60
Speed (m/s)	0	15	24	32	40	45	47

2 The graph shows the rate at which a cup of tea cools over time.

a What temperature is the tea after 15 minutes?

b What temperature would you expect the tea to be 60 minutes after the experiment begins?

c Is your prediction reliable? Explain your answer.

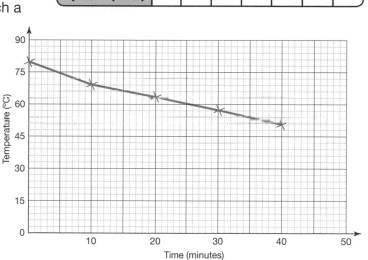

3 The graph shows the number of tourists who visit a holiday destination each year.

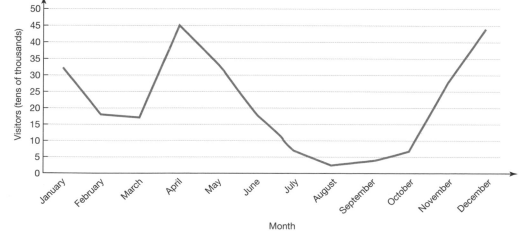

a How many tourists visited in March?

b During which month were there: i the most tourists   ii the fewest tourists?

c Describe the pattern in the number of tourists over the year.

225

Scatter graphs

A scatter graph is a good way to identify relationships between two sets of data.

The table shows the results of 10 students in their science and maths tests.

Maths	73	92	34	19	35	34	74	70	38	82
Science	69	83	50	20	29	100	69	70	44	88

These results are plotted in a scatter diagram:

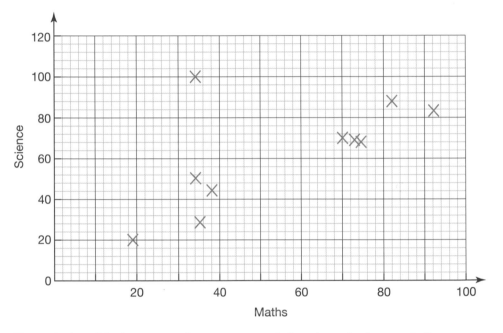

NAILIT!

If you are asked to describe the relationship between two sets of values in a scatter diagram, you need to decide if it is positive, negative or if there is no correlation (see Snap it!). You must also explain what this means.

The relationship between the two sets of data is called **correlation**.

If the plotted points all cluster near to a line, this is called a **strong correlation**. If they are only loosely positioned around a line it is called a **weak correlation**.

SNAPIT! **Correlation**

Positive correlation
As the x value increases, the y value increases.

Negative correlation
As the x value increases, the y value decreases.

No correlation
There is no relationship between the x and y values.

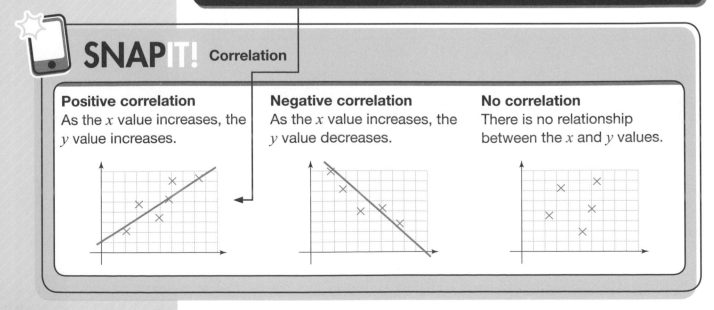

In the example on the previous page the data for the students has a positive correlation. This means that as the maths mark increases, the science mark increases.

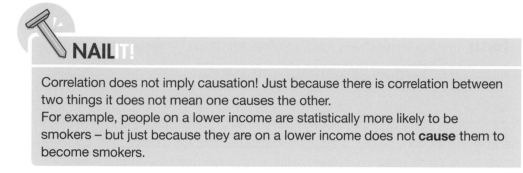

NAILIT!

Correlation does not imply causation! Just because there is correlation between two things it does not mean one causes the other.
For example, people on a lower income are statistically more likely to be smokers – but just because they are on a lower income does not **cause** them to become smokers.

Line of best fit

A line of best fit can be drawn on a scatter graph if there is positive or negative correlation.

Draw a straight line following the trend of the data – you should aim to have the same number of points on each side – but ignore any outliers.

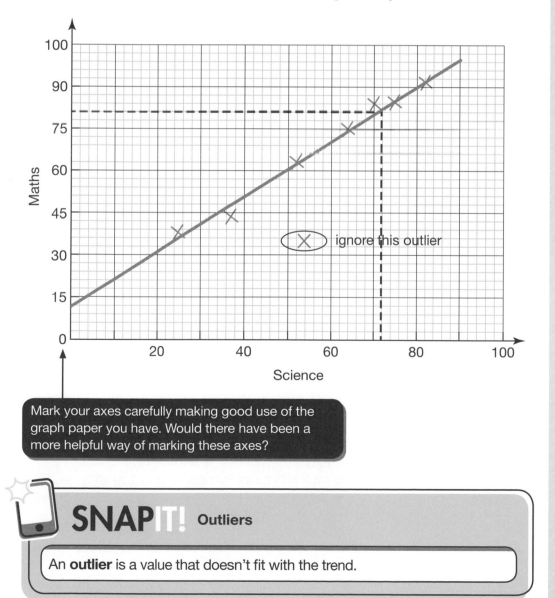

ignore this outlier

Mark your axes carefully making good use of the graph paper you have. Would there have been a more helpful way of marking these axes?

SNAPIT! **Outliers**

An **outlier** is a value that doesn't fit with the trend.

The outlier is ignored – this value may be caused by a student being unwell on the day of the test or other influences, so it should not be allowed to affect the trend of the data.

The line of best fit can be used to predict missing values.

NAIL IT!

The line of best fit does **not** have to go through the origin.

For example, you could predict the mark that a student who scored 72% in the science exam is likely to achieve in maths. Draw a vertical line up to the graph from 72 on the x-axis, then read across horizontally to the y-axis – giving 81%.

Reading data values between the lowest and highest values of a data set is called **interpolation** – it is quite reliable because it is within the recorded data.

In the example above, you can confidently predict that the student's maths mark is likely to be around 81%.

Taking a data value from outside the data set is called **extrapolation** – it is less reliable since you do not know that the data will continue to follow this trend beyond the highest and lowest values recorded.

If the student had scored 15% in science, this value is lower than the lowest science score – so reading the maths mark from the graph will be less reliable. It doesn't mean you can't do this – but any answers found using extrapolation should be treated with caution!

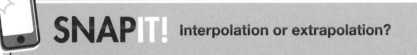

SNAP IT!   Interpolation or extrapolation?

Interpolation
Values read from within the data set – these are fairly reliable

Extrapolation
Values read from outside the data set – these are less reliable

NAIL IT!

Remember any values you read from the line of best fit will only be estimates since your line of best fit may be slightly different from those drawn by other people.

WORKIT!

The scatter graph shows the average daily temperature plotted against the number of heaters sold by a department store.

a Describe the type of correlation and explain what this means.

Negative correlation – the higher the average temperature, the fewer heaters are sold.

b Explain why you would not use a line of best fit to predict how many heaters you would sell on a day when the average temperature was 15°C.

15°C lies outside the range of the data so you would be extrapolating, therefore the prediction would be unreliable.

CHECKIT!

1 The scatter diagram shows the length and weight of eight babies born in a hospital.

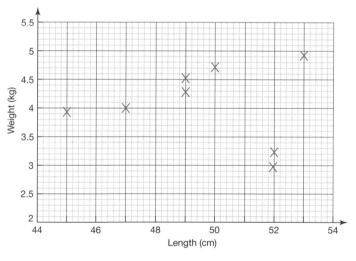

a Draw a ring round any outliers.

b Draw a line of best fit on the diagram.

c What type of correlation is displayed in the diagram?

d Use your line of best fit to estimate the weight of a baby measuring 51 cm.

e Explain why you wouldn't use your line of best fit to predict the length of a baby weighing 5.5 kg.

2 In an experiment seedlings are grown different distances from a light source. The distance from the light source and height of the seedling after 2 weeks are recorded.

Distance from light source (cm)	Height (cm)
0	4.2
25	3.9
50	3.8
75	2.5
100	3.0
125	0
150	2.2

a Plot a scatter graph to show the data.

b Draw a line of best fit on the data.

c Draw a ring round any values you consider to be outliers and explain why these might have occurred.

d Copy and complete the sentence:

The further the seedling is from the light source, the...

3 An economics researcher says, 'There is positive correlation between the number of betting shops in an area and the crime rate.'

Do you think this correlation implies causation? (That is, does the number of betting shops cause the crime rate to be higher?) Explain your answer.

Graphical misrepresentation

Sometimes graphs and charts are drawn deliberately inaccurately in order to mislead the reader. The examples below show some ways this is done.

Pictogram

In a pictogram, the symbols used should all be the same, and should represent exactly the same value. In the pictogram illustrated, a secondary school is trying to promote its reputation for rugby. The pictogram suggests rugby is far more popular than football, but look at the key!

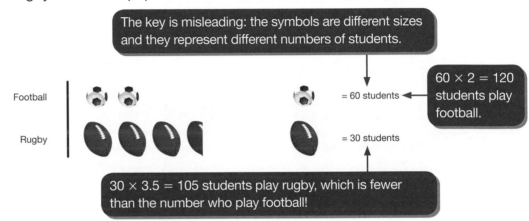

The key is misleading: the symbols are different sizes and they represent different numbers of students.

$60 \times 2 = 120$ students play football.

= 60 students

= 30 students

$30 \times 3.5 = 105$ students play rugby, which is fewer than the number who play football!

DO IT!

When you see a graph or chart, not just in an exam but in real life too, look at it critically.

Is it correctly plotted and labelled? Have they designed it to give you helpful information?

Maybe they are hiding something, or exaggerating something – try to work out if this is the case.

Bar chart

This bar chart shows the number of people in a certain country who care, or don't care, about the loss of wildlife habitats. It looks like there are a great majority who don't care, but look carefully at the *y*-axis!

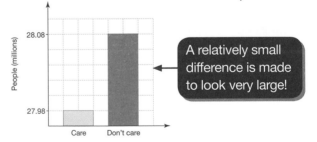

A relatively small difference is made to look very large!

Time series graph

Here is a graph showing a company's percentage profit over five years. It seems to be growing extremely rapidly. However, the axes are not marked evenly. This incorrectly exaggerates the growth in profit.

NAIL IT!

Always check the axes carefully. Are they evenly labelled?

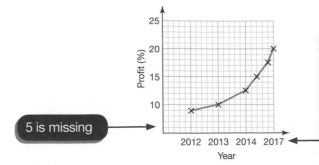

5 is missing

Look closely at the axis markings. You can see that they are not regular, so it looks as if the profits have risen sharply in the last 3 years.

The graph below left shows the Earth's surface temperature since 1880, plotted using a mean value for every five years of data. The vertical axis on the left has a zigzag to show that part of the scale is missing. This enables the trend to be seen more clearly, and it is evident that temperatures have been rising.

An organisation that wants to suggest temperatures are not rising publishes a different graph (below right). By choosing a narrow range of data and an annual, not five-year, average (mean), they suggest that although temperatures vary they are not generally going up. This graph also misses out the average for 2010 (which was nearer to 14.6°C). So, although it is correctly plotted, the second graph is quite misleading.

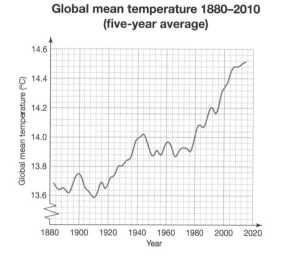

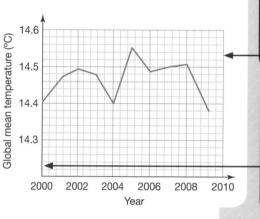

An important value for 2010 has been missed off. No more recent data is provided.

There is no zigzag to represent the missing axis values, so the data plotted looks more significant than it is.

CHECKIT!

1 A political lobbying organisation are claiming that global warming is not happening. On their website they use the following excerpt from a graph to support their argument.

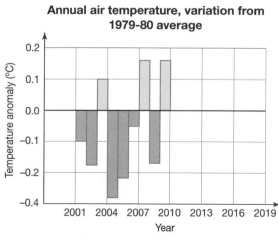

Explain why this data might be misleading.

2 The three bar charts below all show the same information about 80 users of a squash court at a gym. Which bar chart accurately displays the information? Explain why the others are inaccurate.

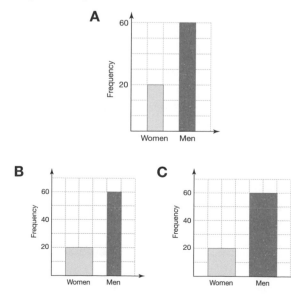

🔲 You may **not** use a calculator for these questions.

1 Henry carries out a survey about people's favourite TV programme. He asks 10 of his friends.

Give two reasons why his sample is biased.

2 India carried out a survey of 20 students to find out what their favourite pizza toppings were.

The table gives information about her results.

Topping	Frequency
Margherita	11
Vegetarian	2
Hawaiian	6
Pepperoni	1

a Which topping is the mode?

b In India's survey, what percentage of people like pepperoni best?

c India is going to draw a pie chart to display her information. What angle should the sector for pepperoni be?

3 The pie chart below shows the colour of electric cars sold by a garage one year.

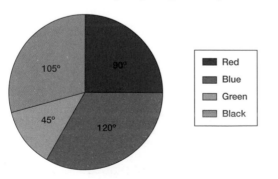

Red
Blue
Green
Black

a What fraction of the cars sold were red?

60 green cars were sold.

b How many cars does the whole pie chart represent?

c How many black cars were sold?

4 The graph gives information about the percentage of people who do their grocery shopping online.

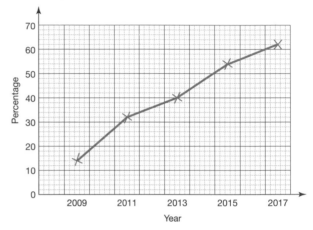

a Describe the trend in the percentage of people who do their grocery shopping online.

b Use the graph to predict the percentage of people who will shop online in 2019.

c Is your predication reliable? Explain your answer.

5 The stem and leaf diagram gives information about the height of seedlings grown in two different locations.

```
Greenhouse        Outside
            2 | 0 1 1 3 8 9
      9 9 7 | 3 | 1 1 5
  8 7 7 7 3 | 4 | 1
      1 0 | 5 |
```

Key	Key
Greenhouse:	Outside:
7 | 3 = 3.7 cm	2 | 0 = 2.0 cm

a For the seedlings grown both inside the greenhouse and outside, work out the:
   i mode **ii** median **iii** range.

b Use your answers to compare the two sets of data, describing any differences you notice.

c What is the range of the height of all the seedlings?

6 The table shows some information about the pets owned by students.

	Cat	Dog	Fish	Rabbit
Girls	3	5	1	0
Boys	8	2	2	4

a Draw a suitable diagram or chart for this information.

b What fraction of students have a cat?

7 The table shows information about the age of customers using a library.

Age	Frequency
0–19	17
20–39	2
40–59	32
60–79	23
80–99	9

a What is the median class?

b Alin calculates the range of ages by working out 99 − 0 = 99
The range may not be 99. Explain why.

8 A shoe shop records the size of shoes sold one morning. The information is shown in the bar chart.

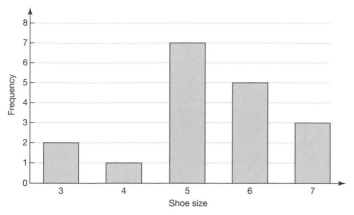

a How many size 5 shoes were sold?

b What is the mode of the sizes of shoe sold?

c Calculate the mean size of shoe sold.

d Which measure of central tendency is most useful for the manager to know? Give a reason for your answer.

9 An athlete records the times (in seconds) she takes to run 800 m in ten training runs.

112 115 121 118 132

129 131 120 112 119

a Show the information in a stem and leaf diagram.

b What proportion of her runs are under 120 seconds?

10 A further education college has drawn up a timetable for the three science A levels. There are 150 places. The pictogram shows how many places there are for each science.

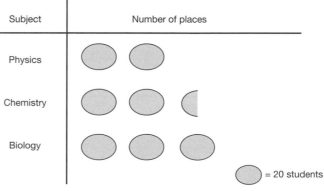

a What fraction of the places were for studying chemistry?

b How many more places were allocated for biology than physics?

c What is the modal subject studied?

11 The ages of 9 people and the number of minutes they spend using a mobile phone each day are recorded.

Age	18	23	45	17	32	65	56	36	40
Minutes	62	54	15	60	25	13	67	28	16

a Draw a scatter graph to show the data.

b What type of correlation is shown?

c Put a ring around any outliers.

d Use your line of best fit to predict how many minutes a 30 year old would spend on the phone.

e Explain why you would be unlikely to use this graph to predict how long a 70 year old would spend on the phone.

f Describe in words the relationship between the time spent on the phone and the age of a customer.

⊞ You **may** use a calculator for these questions.

12 The table shows the weight of 10 patients at a doctors' surgery.

Weight, w (kg)	Frequency
$60 \leq w < 70$	3
$70 \leq w < 80$	5
$80 \leq w < 90$	2

a Calculate an estimate of the mean weight.

b Explain why this is only an estimate.

13 In a local election there are 36 579 voters.

The voters are grouped into four categories using their age and gender.

Voter category	Number of voters
Male < 50	12 201
Female < 50	10 678
Male ≥ 50	5699
Female ≥ 50	8001

A stratified sample of 120 people is taken from a population.

How many of each category should be included?

14 A sample of 5000 people are surveyed about their annual income.

Annual income, x	Percentage of population
$0 \leq x < £25\,000$	35%
$£25\,000 \leq x < £40\,000$	55%
$£40\,000 \leq x < £80\,000$	10%

Draw a pie chart to represent the data.

15 An athlete records the length of time it takes her to run 100 m during 10 different training sessions. She works out her mean time is 10.3 seconds.

The following day she runs 100 m in 9.5 seconds. What is her new mean time?

16 Catherine rolls a dice 5 times. Her mean score is 3.8. The range of scores is 4. The modal score is 3.

What numbers did she roll?

Answers

For full worked solutions, visit:
www.scholastic.co.uk/gcse

Number

Basic number techniques

1 a false   **c** true   **e** true
   b true   **d** true

2 −0.3, −1.5, −2.5, −4.2, −7.2

3 0.049, 0.124, 0.412, 0.442, 1.002

4 a <   **b** <   **c** >

Factors, multiples and primes

1 a 5   **b** 1, 12
   c 1, 5, 45   **d** 1, 5

2 HCF = 10, LCM = 1050

3 $2 \times 3^2 \times 5$

4 a 10   **b** 840

5 12 and 18

Calculating with negative numbers

Stretch it! negative, yes

1 a −11   **c** −6   **e** 0
   b 99   **d** 18   **f** 25

2 −8 and 9

3 32°C

Division and multiplication

Stretch it! 148 419

1 a 2115   **b** 56 364

2 a 47   **c** 126 remainder 4 or $126\frac{4}{17}$
   b 516

3 a 33 boxes   **b** 1 pencil

4 £91.25

5 £288

6 $307\frac{2}{3}$

7 28 805

8 a 682   **b** 13   **c** 5

9 37 boxes

10 He has not placed a zero in the units column before multiplying through by 5.

Calculating with decimals

Stretch it! 18.2

1 a 2.33   **c** 0.035   **e** 1.563   **g** −6
   b 24.391   **d** 6.099   **f** 0.6   **h** −5.04

2 £4.64

3 Erica: £54.92; Freya: £27.46

Rounding and estimation

Stretch it! a 1.0   **b** 1.00   **c** 1.000 − they are all 1

Stretch it! 55.25 m² − an overestimate.

1 a 0.35   **c** 32.6
   b 10   **d** 33 100

2 a $150 \le x < 250$ **c** $3.15 \le x < 3.25$
   b $5.5 \le x < 6.5$ **d** $5.055 \le x < 5.065$

3 $\frac{30}{0.5 \times 6} = 10$

4 a 23 580   **c** 23 600   **e** 20 000
   b 23 580   **d** 24 000

5 b is false since $18 \times 1 = 18$ so 18×0.9 cannot be 1.62

   c is false since if you divide by a number smaller than 1 the answer will be larger.

6 Tarik should choose One tariff.

Converting between fractions, decimals and percentages

Stretch it! $0.\dot{1}, 0.\dot{2}, 0.\dot{3}, \ldots 0.\dot{4}, 0.\dot{5}$

1 a $\frac{32}{100} = \frac{8}{25}$ **c** $\frac{33}{100}$
   b $1\frac{24}{100} = 1\frac{6}{25}$ **d** $\frac{95}{100} = \frac{19}{20}$

2 a 0.416   **c** 0.4̇9̇   **e** 0.4̇28571̇
   b 0.375   **d** 0.185

3 a 91%   **c** 80%
   b 30%   **d** 60%

4 37.5%

5 30%, 0.35, $\frac{2}{5}$

6 $\frac{15}{20} = \frac{75}{100} = 75\%$ – Amy

   Rudi was highest

Ordering fractions, decimals and percentages

1 $\frac{7}{12}, \frac{3}{8}, \frac{1}{3}$

2 $-2.2, -\frac{1}{10}, 1\%, 0.1, 15\%, \frac{1}{5}, 7$ (so the middle is 0.1)

3 Yes, if the numerator of a fraction is half the denominator then the fraction is equivalent to $\frac{1}{2}$. If the numerator is smaller than this the fraction must be less than $\frac{1}{2}$.

Calculating with fractions

Stretch it! No, you could add the whole number parts then the fraction parts, giving:

$1 + 2 = 3$

$\frac{3}{5} + \frac{1}{4} = \frac{17}{20}$

$\qquad = 3\frac{17}{20}$

1 a $1\frac{5}{8}$ **c** $\frac{10}{21}$ **e** $\frac{2}{25}$
   b $\frac{6}{17}$ **d** $8\frac{3}{20}$

2 a 12   **b** £35   **c** 808 mm

3 20

4 35

Percentages

1 a 1.8 cm   **b** £0.30   **c** 4 ml

2 a 33   **b** 540   **c** £101.92

3 a 480   **b** 133   **c** £14.58

4 3052

5 £14 300

Order of operations

1 a 7 **b** −1.9 **c** −13

2 30

3 $(8 − 3 + 5) × 4$

Exact solutions

1 a π **b** 36π **c** $2\frac{1}{2}\pi$

2 a 7π **b** $\frac{5}{8}\pi$

3 area = $\frac{6}{28} = \frac{3}{14}$ cm²

perimeter = $2\frac{1}{14}$ cm

4 a 18π cm **b** 144π cm²

5 $\frac{1}{2}\pi$ cm

Indices and roots

1 a $\frac{1}{3}$ **b** $2\frac{1}{2}$ **c** $1\frac{1}{9}$

2 $\frac{1}{12}, 1^3, \sqrt[3]{8}, \sqrt[3]{27}, 3.7, 3^2$

3 a −8 **c** 81

b 1 **d** 1

4 a $\frac{1}{4}$ **b** $\frac{1}{49}$ **c** 1 **d** $\frac{1}{3}$

5 5^4

6 64

Standard form

1 a 45 000 000 **b** 0.091

2 a $6.45 × 10^8$ **b** $7.9 × 10^{-8}$

3 345 800

4 $3.1 × 10^{-2}$ $3.09 × 10$ $3 + (2.1 × 10^2)$ $3.2 × 10^2$

5 $3 × 10^8$

6 $0.022 = 2.2 × 10^{-2}$ m = 2.2 cm

Listing strategies

Stretch it!

red + small, red + medium, red + large,
green + small, green + medium, green + large,
blue + small, blue + medium, blue + large.

1 111
112, 121, 211, 113, 131, 311

222
221, 212, 122, 223, 232, 322

333
331 313 133 332 323 233

123 132 213 231 312 321

2 444 446 449
464 466 469
494 496 499

3 Small A, Small B, Small C, Small D
Medium A, Medium B, Medium C, Medium D
Large A, Large B, Large C, Large D.

Review it!

1 13294

2 7 and 6 (or 11 and 2, where both are prime and 2 is also a factor of 12)

3 $620 = 2 × 2 × 5 × 31 = 2^2 × 5 × 31$

4 $18 = 2 × 3 × 3$
$36 = 2 × 2 × 3 × 3$
$40 = 2 × 2 × 2 × 5$
HCF = 2

5 −11.5, −8.3, −3.5, −3.2, 1.4

6 a £51.73 **b** £18.33

7 a 3.22 **c** −0.12 **e** −4
b 4023 **d** −4.8

8 $30 × \sqrt{16} + 17 = ± 120 + 17 = 137$ or −103

9 27

10 $31\frac{4}{11}$

11 a 0.375 **b** 70%

12 a $\frac{7}{10}$ **b** $\frac{4}{5}$

13 All of them

14 a $\frac{26}{35}$ **b** $1\frac{1}{2}$ **c** $1\frac{1}{2}$

15 0.25 − 0.07

16 $\frac{3}{4}$

17 $\frac{9}{200}$

18 £2.15

19 a 9 **b** 5

20 a $3.4 × 10^9$ **b** $3.04 × 10^{-7}$

21 $37.55 \leq x < 37.65$

22 a 51
b 12, 15, 21, 51, 25, 52

23 a $200 × 9 × 10 = 18 000 = £180.00$
b Underestimate since all numbers were rounded down.

24 240

25 35%

26 no since 2 is a prime number and odd + odd + even = even

27 £279.20

28 a 3.1 **b** 3.05

29 a 325 000 **b** 320 000

30 729

31 £16.62

32 3420

33 £112.53

Algebra

Understanding expressions, equations, formulae and identities

1 a $3a + 6 = 10$ **c** $3(a + 2)$
b $C = \pi D$ **d** $3ab + 2ab = 5ab$

2 James is correct.

$4x − 2 = 2x$ can be solved to find the value of x so it is an equation.

Or, the two sides of $4x − 2 = 2x$ are not equal for all values of x so it cannot be an identity.

Simplifying expressions

Stretch it! $12t \times t \times t$, $2t \times 6t \times t$, $2t \times 3t \times 2t$, $3t \times 4t \times t$

1 a p^3 c $12ab$ e $-8g^2$

   b $28bc$ d $20x^2$ f $6pqr$

2 a $5x$ c 6 e $4x$

   b $-7w$ d $4n$ f $9a$

Collecting like terms

1 a $5f$ h $-3a - b - 3$

   b $7b$ i $2x^2$

   c $5mn$ j t^3

   d $3a + 1$ k $2a + b^2$

   e $4d - 2e$ l $7\sqrt{x}$

   f $5x + 3y - 2$ m $3\sqrt{x}$

   g $7a + 5b$ n $7\sqrt{x}$

Using indices

1 a x^9 c $6m^8$ e u^3

   b p^5 d $15m^8n^4$ f t

2 a x^2 d $2x^3$ g $\frac{x}{3}$

   b y^4 e $\frac{1}{m^2}$

   c p f $\frac{x^4}{3}$

3 a x^6 c p^{10} e $\frac{1}{x^6}$

   b y^{16} d $16m^{10}$ f n^8

4 a $12x^3$ c $\frac{1}{y^2}$

   b $5x^3$ d a^5b^3

5 x^4

Expanding brackets

Stretch it!

a $\sqrt{3}a + a^2$ c $c + d$

b $\sqrt{5}b - b^2$

Stretch it!

1 $(x + 2)(x + 4) = x^2 + 6x + 8$

2 a $2x^2 + 8x + 6$ d $x^2 + xy - 2y^2$

   b $3x^2 + 10x - 8$ e $6x^2 - xy - y^2$

   c $6x^2 + 7x - 3$

1 a $3a + 6$ f $-2y - 4$

   b $4b - 16$ g $x^2 - 2x$

   c $10c + 25$ h $2a^2 + 10a$

   d $6 - 2e$ i $3x + 6y$

   e $4x + 4y + 8$ j $-2a + 2b$

2 a $3a - 5$ b $6x + 4$

3 a $8x + 26$ b $17y - 3$ c $2m + 31$

4 a $x^2 + 5x + 6$ c $a^2 - 4a - 21$

   b $y^2 + y - 12$ d $m^2 - 7m + 6$

5 a $x^2 + 2x + 1$ c $m^2 - 4m + 4$

   b $x^2 - 2x + 1$ d $y^2 + 6y + 9$

Factorising

Stretch it! $(x + 1)$

Stretch it!

a $(a + \sqrt{3})(a - \sqrt{3})$ b $(b + \sqrt{5})(b - \sqrt{5})$

1 a $3(a + 3)$ c $7(1 + 2c)$

   b $5(b - 2)$ d $d(d - 2)$

2 a $4(2a + 5)$ c $9(2 + c)$

   b $4(b - 3)$ d $d(2d - 3)$

3 a $2(2x - 3y)$ e $n(2 - 9n)$

   b $m(a + b)$ f $5x(1 + 2y)$

   c $4a(3a + 2)$ g $4p(q - 3)$

   d $x(4x + 3y)$ h $4y(x^2 - 2)$

4 $a = 2$, $b = 3$

5 a $(x + 1)(x + 7)$ e $(x - 3)(x - 3)$

   b $(x - 1)(x + 5)$ f $(x + 3)(x + 4)$

   c $(x + 2)(x - 4)$ g $(x - 2)(x + 5)$

   d $(x - 2)(x - 3)$ h $(x + 4)(x - 5)$

6 a $(x + 4)(x - 4)$ c $(x + 9)(x - 9)$

   b $(x + 6)(x - 6)$ d $(y + 10)(y - 10)$

Substituting into expressions

1 11

2 a 10 c -22 e 28

   b -24 d 20 f -1

3 False.

   When $a = 3$: $3a^2 = 3 \times 3^2 = 3 \times 9 = 27$.

4 a -20 c -8

   b 2 d 26

5 -3.5

Writing expressions

Stretch it!

a $8x + 4$ b $2x + 1$

1 a $4 - q$ c xy (or yx)

   b $n + m$ (or $m + n$)   d p^2

2 $x + y$

3 £$\frac{y}{8}$

4 $100n + 75b$

5 $9a + 2$

6 $4x + 10$

Solving linear equations

1 a 7 c 16

   b 17 d -2

2 a 5 c -4 e 11

   b 5 d 33 f 19

3 a 2 c $\frac{5}{2}$ (Or 2.5, or $2\frac{1}{2}$)

   b 3 d -2

4 Hannah has not subtracted 4 from *both* sides.

5 a 3 d -4

   b 3 e $-\frac{4}{8} = -\frac{1}{2}$ (Or -0.5)

   c 3

2 3 2 2 2 2 2 2 2 2 2 2 2 2 2 2 2 2 2 2 2 2 2 2 2 2 2 2 2 2 2 2 2 2

Answers

6 **a** 3 **d** 3
  b 3 **e** −6
  c $\frac{5}{2}$ (Or 2.5 or $2\frac{1}{2}$)
7 **a** −4 **d** 1
  b $\frac{9}{6} = \frac{3}{2}$ (or 1.5) **e** $\frac{9}{6} = \frac{3}{2}$ (Or 1.5 or $1\frac{1}{2}$)
  c 2

Writing linear equations

Stretch it!   10.5 cm

1 **a** $8s + 12$ **b** 9 cm
2 **a** 50 **b** 115°
3 30 years old
4 4
5 8 cm
6 $a = 18, b = 28$

Linear inequalities

1 **a** $x = 3, 4, 5$ **c** $x = 0, 1, 2, 3$
  b $x = 2, 3, 4, 5$ **d** $x = -3, -2, -1, 0, 1$
2 **a** $x < 3$ **c** $-1 \le x \le 5$
  b $x \ge -2$
3 **a**
  b
  c
  d
4 **a** $x > 3$
  b $x \le \frac{10}{4} = \frac{5}{2}$
  (Or $x \le 2.5$ or $x \le 2\frac{1}{2}$)
  c $x < -5$
  d $x \ge -1$
5 Olivia has not multiplied all the terms in the bracket by the term outside.
6 **a** $-2, -1, 0, 1, 2$
  b $-4, -3, -2, -1, 0, 1, 2, 3, 4, 5, 6$
  c $0, 1, 2, 3$
  d $3, 4, 5$
7 **a** $-\frac{5}{2} < x$ (Or $x > -\frac{5}{2}$) **b** −2
8 **a** $3 \le x$ (Or $x \ge 3$)
  b $-1 > x$ (Or $x < -1$)
  c $\frac{1}{2} \ge x$ (Or $x \le \frac{1}{2}$)
  d $2 > x \ge -3$

Formulae

Stretch it!
$4a = 5(2b^2 - a)$
$4a = 10b^2 - 5a$
$9a = 10b^2$
$\frac{9a}{10} = b^2$
$\sqrt{\frac{9a}{10}} = b$

1 £305
2 27
3 113°F
4 −90
5 1.5
6 $C = 25d + 50$
7 **a** $k = \frac{8m}{5}$ **b** 320 km
8 $A = l^2$
9 **a** $4a + 6$ **b** 30 cm
10 −65
11 **a** $\frac{v - u}{t} = a$ **d** $\frac{v^2 - u^2}{2a} = s$
  b $\frac{3V}{A} = h$ **e** $t = \sqrt{\frac{2s}{a}}$
  c $\frac{y + 9}{3} = x$ **f** $g = \frac{2s}{T^2}$
  or: $x = \frac{1}{3}y + 3$ **g** $a = \frac{3}{(4x - 1)}$

Linear sequences

1 **a** **i** 14, 17 **ii** 7, 3 **iii** 27, 33 **iv** 24, 29
  b **i** 29 **ii** −13 **iii** 57 **iv** 49
2 **a** 1st term = 2, 2nd term = 6, 3rd term = 10, 4th term = 14
  b 78
3 **a** 3, 10 **b** 8th term
4 **a**   **b** 16
  c No. The number of triangles forms an even number sequence and 35 is odd.
5 **a** $4n - 1$
  b $4n - 1 = 99 \ (+ 1)$
  $4n = 100 \ (\div 4)$
  $n = 25$
  Yes, 99 is a term in the sequence because 25 is an integer.

Non-linear sequences

1 1, 3, 5, 7, 9, … Arithmetic sequence
  1, 2, 4, 8, 16, … Geometric sequence
  1, 4, 5, 9, 14, … Fibonacci-type sequence
  1, 4, 9, 16, 25, … Square-number sequence
2 27, 38
3 **a** $\frac{1}{2}, \frac{1}{4}$ **d** 3, 9
  b 0.005, 0.0005 **e** −0.8, −1.6
  c $\frac{1}{16}, \frac{1}{32}$ **f** −24, 48

4 42

5 6, 9, 14, 21

6 71

7 3, 8, 15

8 a 4th term $= b + a + b = a + 2b$

5th term $= a + b + a + 2b = 2a + 3b$

b 4

Show that...

1 a $3 + 5 = 8$ (or any other two primes except 2)

b Mo is not correct.

If x is even, then $x + 1$ is odd, and $2(x + 1)$ is even. Substituting into $\frac{n}{2}$ gives $\frac{2(x + 1)}{2} = x + 1$, an odd number as stated above. Therefore $\frac{n}{2}$ is not always even when n is even.

Or:

n could have more than one even factor.

For example,

12 (even) $= 2 \times 2 \times 3$.

$\frac{12}{2} = 6$ (even)

2 a LHS $= (x + 2)(x - 2) \equiv x^2 - 2x + 2x - 4 = x^2 - 4$

RHS $= x^2 - 4$

LHS $\equiv$ RHS

So $(x + 2)(x - 2) \equiv x^2 - 4$

b LHS $= (x - 3)^2 \equiv (x - 3)(x - 3)$

$= x^2 - 6x + 9$

RHS $= x^2 - 6x + 9$

LHS $\equiv$ RHS

So $(x - 3)^2 \equiv x^2 - 6x + 9$

c LHS $= (x + 1)^2 + 4 \equiv (x + 1)(x + 1) + 4$

$= x^2 + 2x + 1 + 4 = x^2 + 2x + 5$

RHS $= x^2 + 2x + 5$

LHS $\equiv$ RHS

So $(x + 1)^2 + 4 \equiv x^2 + 2x + 5$

d LHS $= 6(a - 3) - 2(2a - 5) + 6$

$= 6a - 18 - 4a + 10 + 6 = 2a - 2$

RHS $= 2(a - 1) = 2a - 2$

LHS $\equiv$ RHS

So $6(a - 3) - 2(2a - 5) + 6 \equiv 2(a - 1)$

e LHS $= 4(x - 3) + 2(x + 5) = 4x - 12 + 2x + 10$

$= 6x - 2$

RHS $= 3(2x - 1) + 1 = 6x - 3 + 1 = 6x - 2$

LHS $\equiv$ RHS

So $4(x - 3) + 2(x + 5) \equiv 3(2x - 1) + 1$

3 $4(ax - 2) + 5(3x + b) \equiv 23x - 3$

LHS $= 4ax - 8 + 15x + 5b = (4a + 15)x + (5b - 8)$

Given that RHS $\equiv$ LHS,

$4a + 15 = 23$

$4a = 8$

$a = 2$

and $5b - 8 = -3$

$5b = 5$

$b = 1$

4 Rod A $= n$

Rod B $= n + 1$

Rod C $= n + 2$

Rod A + Rod C $= n + n + 2$

$= 2n + 2$ (Factorise)

$= 2(n + 1)$

Rod A + Rod C is 2 times the length of rod B.

Functions

1 a 0

b -2

c -12

2 a

$\times 1 \rightarrow -2$	
x	y
-2	0
0	2
2	4

b

$\times 2 \rightarrow -3$	
x	y
-2	-7
0	-3
2	1

3 a $y = 3x - 3$

b 27

c $x = \frac{y + 3}{3}$

d If $x = y$ then $x = 3x - 3$

$2x = 3$, $x = \frac{3}{2}$ (or 1.5)

Substituting $x = 1.5$ into the function,

$y = 1.5 \times 3 - 3 = 1.5 = \frac{3}{2}$.

So x and y can be equal.

Coordinates and midpoints

Stretch it!

$(-4, 2)$

1 a A(2, 2)

b and **c**

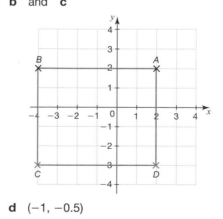

d $(-1, -0.5)$

Straight-line graphs

Stretch it! $x = 0.5$

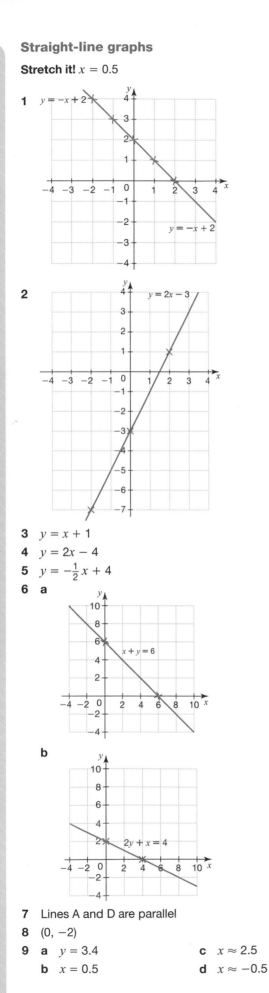

1 $y = -x + 2$

2 $y = 2x - 3$

3 $y = x + 1$

4 $y = 2x - 4$

5 $y = -\frac{1}{2}x + 4$

6 **a** $x + y = 6$

   b $2y + x = 4$

7 Lines A and D are parallel

8 $(0, -2)$

9 **a** $y = 3.4$ **c** $x \approx 2.5$

   b $x = 0.5$ **d** $x \approx -0.5$

Solving simultaneous equations

1 $x = \frac{21}{2}$ (or 10.5 or $10\frac{1}{2}$)
   $y = \frac{11}{2}$ (or 5.5 or $5\frac{1}{2}$)

2 **a** $x = 1, y = 2$ **d** $x = 3, y = 2$

   b $x = 3, y = -2$ **e** $x = 4, y = -1$

   c $x = 6, y = -4$ **f** $x = 2, y = 1$

3 7 and 14

4 A burger costs 95 p. A cola costs £1.10.

5 Adult ticket costs £25, Child ticket costs £2.50

6 **a** $x = 2, y = 8$ **b** $x = 8, y = 1$

Quadratic graphs

Stretch it! $x \approx 3.8$ and $x \approx -0.8$

1 **a**

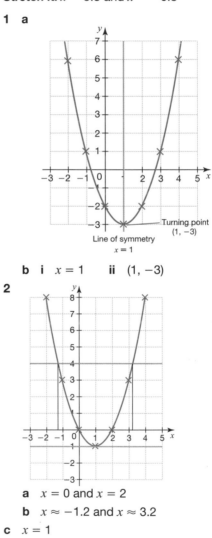

Turning point
$(1, -3)$

Line of symmetry
$x = 1$

   b i $x = 1$ **ii** $(1, -3)$

2

   a $x = 0$ and $x = 2$

   b $x \approx -1.2$ and $x \approx 3.2$

   c $x = 1$

3 a and b

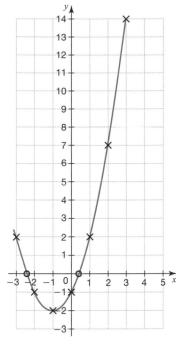

b −2.4 and 0.4

Solving quadratic equations

Stretch it!

$x = 4$ or $x = -4$

$x = 5$ or $x = -5$

1 a $x = 0$ or $x = 4$ **d** $x = -1$ or $x = -9$

  b $x = 0$ or $x = -7$ **e** $x = 3$ or $x = -4$

  c $x = 4$ or $x = -4$ **f** $x = -2$ or $x = 8$

2 a $x = -7$ or $x = 7$

  b $x = 0$ or $x = 3$

  c $x = -1$ or $x = -6$

Cubic and reciprocal graphs

1 a

x	−3	−2	−1	0	1	2	3
y	−28	−9	−2	−1	0	7	26

b

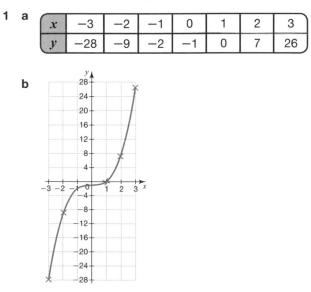

2 a

x	−4	−3	−2	−1	$-\frac{1}{2}$	$-\frac{1}{4}$	$\frac{1}{4}$	$\frac{1}{2}$	1	2	3	4
y	$\frac{1}{4}$	$\frac{1}{3}$	$\frac{1}{2}$	1	2	4	−4	−2	−1	$-\frac{1}{2}$	$-\frac{1}{3}$	$-\frac{1}{4}$

b

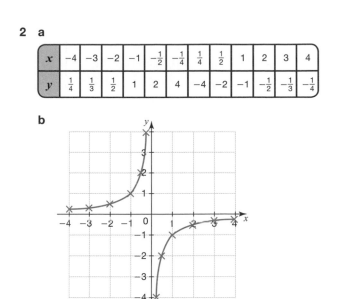

Drawing and interpreting real-life graphs

1 a

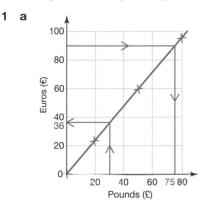

b The graph is a straight line with a positive gradient. As the number of pounds steadily increases, the corresponding number of euros steadily increases. This is direct proportion.

c i €36   **ii** £75

d From the graph: £30 = €36

So £90 = €36 × 3 = €108

The ring is cheaper in France.

2 a £10   **b** 13p

3 a

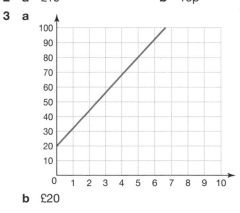

b £20

c Let $d = 6$ days (more than 5 days)

First shop: $C = 12 \times 6 + 20 = 92$

Second shop: $C = 10 \times 6 + 30 = 90$

To hire the sander for more than 5 days use the second shop as it is cheaper.

4 a 30 minutes

b 55 km

c Speed before break = 30 km/hr

Speed after break = 16.7 km/hr

d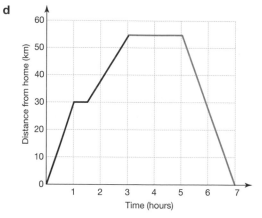

5 a 12.8 m

b 3 seconds

c 0.4 seconds and 2.5 seconds

d The ball is thrown from a height of 3 m above the ground.

6 a 6 m/s **c** 6 seconds

b 4 seconds **d** 1.5 m/s^2

7 a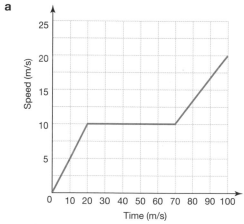

b $0.\dot{3}$ m/s^2

8 a 35 cm

b The person stays in the bath.

c The person got out of the bath.

d Running water into the bath was quicker. The slope of the line between O and A (filling the bath) is steeper than the slope of the line between E and F (emptying the bath).

Review it!

1 a (3, 5)

b, c and e

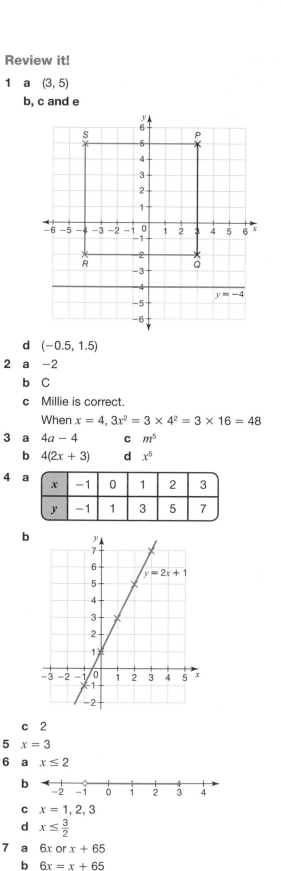

d (−0.5, 1.5)

2 a −2

b C

c Millie is correct.

When $x = 4$, $3x^2 = 3 \times 4^2 = 3 \times 16 = 48$

3 a $4a - 4$ **c** m^5

b $4(2x + 3)$ **d** x^5

4 a

x	−1	0	1	2	3
y	−1	1	3	5	7

b

c 2

5 $x = 3$

6 a $x \le 2$

b

c $x = 1, 2, 3$

d $x \le \dfrac{3}{2}$

7 a $6x$ or $x + 65$

b $6x = x + 65$

$5x = 65$

$x = 13$

Luke is 13 years old.

8 a 33

b No. This is a sequence of odd numbers and 44 is even.

c 47

9 $x^2 + 7x + 12$

10 $a - b = 12$

11 $x = 4, y = 2$

12 a $2x(2x + 3)$

   b $(x + 10)(x - 10)$

   c $(x + 3)(x + 6) = 0$

      either $x + 3 = 0$

                $x = -3$

      or x + 6 = 0

                $x = -6$

13 $a = 2, b = 3$

14 a $12m$ **b** $12p^2$ **c** $6x$

15 a $w = 11$ **b** 21   **c** $5a + 8b$

16 $(8, -3)$

17 a $(-5, 4)$

   b Length = 8 units, width = 6 units

18 a

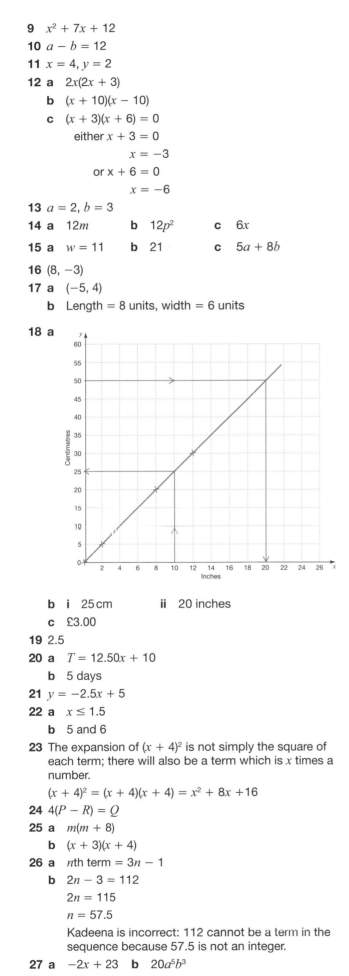

   b i 25 cm   **ii** 20 inches

   c £3.00

19 2.5

20 a $T = 12.50x + 10$

   b 5 days

21 $y = -2.5x + 5$

22 a $x \leq 1.5$

   b 5 and 6

23 The expansion of $(x + 4)^2$ is not simply the square of each term; there will also be a term which is x times a number.

   $(x + 4)^2 = (x + 4)(x + 4) = x^2 + 8x + 16$

24 $4(P - R) = Q$

25 a $m(m + 8)$

   b $(x + 3)(x + 4)$

26 a nth term = $3n - 1$

   b $2n - 3 = 112$

      $2n = 115$

      $n = 57.5$

   Kadeena is incorrect: 112 cannot be a term in the sequence because 57.5 is not an integer.

27 a $-2x + 23$ **b** $20a^5b^3$

28 $x = 4.5$

29 Input = -2

30 $a = 6$

31 a 4th term = $b + a + b = a + 2b$

      5th term = $a + b + a + 2b = 2a + 3b$

      6th term = $a + 2b + 2a + 3b = 3a + 5b$

      7th term = $2a + 3b + 3a + 5b = 5a + 8b$

   b $a = 2, b = 3$

32 C

Ratio, proportion and rates of change

Units of measure

1 a 3000 m   **c** 13 000 cm²   **e** 7200 seconds

   b 75 mins   **d** 3.52 litres   **f** 14 kg

2 4.175 kg or 4175 g

3 $2.2\dot{7}$ kg

Ratio

Stretch it! $\frac{31}{56}$

1 a 1:4   **b** 1:3:4   **c** 4:5

2 7:1

3 Allow 1 part cement for 1.5 parts sand.

4 a 7:1   **b** 100 tickets

5 a 3:2   **b** 120

6 12 cm and 15 cm

7 $s = 20t$

8 a 0.6 kg   **b** 40 g

Scale diagrams and maps

Stretch it! $50 \div x$

1 A, B, F

2 a 36 km   **b** 1.25 cm

3 120 m

4 a 1 km   **b** 250°

Fractions, percentages and proportion

1 $\frac{1}{175}$

2 $\frac{11}{24}$

3 a $\frac{3}{4}$ **b** 25%

4 10%

5 School A: 125:145 = 25:29

   School B: 100:120 = 5:6

   No since the ratios are not equivalent.

6 4.125 g

Direct proportion

Stretch it! Since for two values to be in direct proportion when one is 0 the other must be 0.

1 A and E

2 a i 30 meringues   **b** 7 eggs

    ii 100 meringues

3 2 hours 30 mins

4 A, D

Inverse proportion

1 D

2 22.5 mins

3 $1\frac{2}{3}$ of a day.

4 **a** 2

b The age of the chicken and the number of eggs it lays are in inverse proportion. This means that as the age of the chicken increases so the number of eggs it lays decreases.

Working with percentages

Stretch it! £128

Stretch it! 2% ($1.02^5 \times 100 = £110.41$)

1 **a** £51.50   **b** 992   **c** 12.48

2 12.5%   **5** £150

3 20.9°C   **6** 88.9%

4 25 920

Compound units

Stretch it! $\frac{100}{x}$ mph

1 164 units   **3** 0.24 g/cm³

2 40 minutes   **4** 6 N/m²

5 10.8 km/hour

6 475 gallons per hour (to the nearest whole number)

7 Bolt: 100 m in 9.58 seconds = 10.4 m/s
Cheetah: 120 km/hr = 120000 m/hour = 120000 ÷ 60 m/min = 2000 m/min = 2000 ÷ 60 m/sec = 33.3 m/s The cheetah is fastest.

Review it!

1 **a** 3200 m   **b** 540 seconds   **c** 400 ml

2 4.6 km

3 150 minutes

4 3520 cm² or 0.352 m²

5 30 000 cm²

6 $\frac{5}{12}$

7 13 : 9

8 5 minutes

9 2300 kg/m³

10 25%

11 $\frac{12}{25}$ or 48%

12 $\frac{23}{90}$

13 25 hours = 1 day and 1 hour

14 **a** 30 more   **b** $C = 4J$

15 28%

16 £864

17 6 cm

18 £1591.81

19 8000 people

20 3 hours 9 minutes

21 £15.75

22 20%

23 £3675

24 18 : 13

25 1750 men

26 No — for two things to be in direct proportion when one is zero the other must be zero, the graph does not go through the origin so this is not the case.

27 Neither, since the time taken to cook increases as the weight increases it is not in indirect proportion. It is not in direct proportion since a graph to illustrate the relationship would not go through the origin.

28 11 seconds

29 She is incorrect since the ratio of females to males must be the same for them to have equivalent proportions.

Geometry and measures

Measuring and drawing angles

1 **a** 43°   **b** Acute

2 **a**

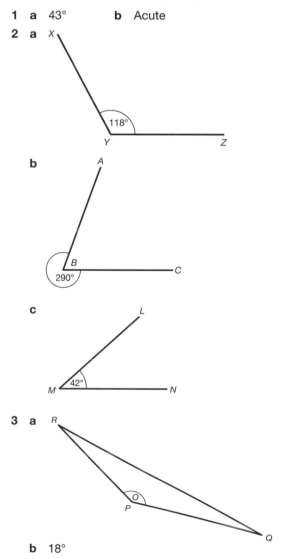

b

c

3 **a**

b 18°

Using the properties of angles

Stretch it! Right angled

1 54°

2 a i 70°

ii Base angles of an isosceles triangle are equal.

b 110°

3 18°

4 a i 54°

ii Angles on a straight line add up to 180°.

b 83°

5 a 96°

b i $y = 96°$

ii Use the fact that corresponding angles are equal, then vertically opposite angles are equal.

Or, use the fact that alternate angles are equal, then use angles on a straight line add up to 180°.

6 a 58°   **b** 64°

c 58° (Alternate angles are equal; or, opposite angles of a parallelogram are equal)

7 Angle $BAD = 62°$ (Opposite angles of a parallelogram are equal)

Angle $ADE = 62°$ (Alternate angles are equal)

$x = 180 − 62 − 62$ (Base angles of an isosceles triangle are equal)

$x = 56°$

8 Angle $ACB = 36°$ (Base angles of an isosceles triangle are equal)

Angle $ABC = 180 − 36 − 36$ (Angles in a triangle add up to 180°)

Angle $ABC = 108°$

$x = 108°$ (Alternate angles are equal)

Using the properties of polygons

Stretch it!

1 The angle sum of a triangle is 180°.

Sum of interior angles of a hexagon = 4 × 180° = 720°.

2

Polygon	Number of sides (n)	Number of triangles formed	Sum of interior angles
Triangle	3	1	180°
Quadrilateral	4	2	360°
Pentagon	5	3	540°
Hexagon	6	4	720°
Heptagon	7	5	900°
Octagon	8	6	1080°
Decagon	10	8	1440°

3 $n − 2$

4 $180° × (n − 2)$

Stretch it!

Angles around a point add up to 360°.

No, 360° is not divisible by 108° (interior angle of a regular pentagon).

1 36°

2 a 24   **b** 3960°

3 Angle $CBE = 72°$

Using bearings

1 a 315°

b

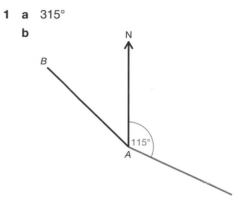

2 344°

3 Kirsty is correct.

The bearing is 314° (360° − 46°) as it must be measured clockwise from north.

Properties of 2D shapes

Stretch it!

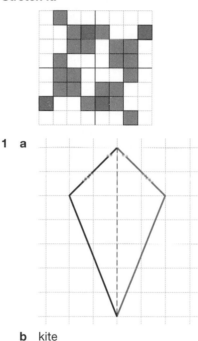

1 a

b kite

2 a 8 possible lines of symmetry:

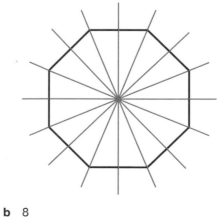

b 8

3 a 2   **c** 1, no

b rhombus   **d** square, rhombus

Congruent shapes

1 Any accurate copy of shape A.

2 a 120° **b** 12 cm

3 a SSS **b** ASA

Constructions

Stretch It!

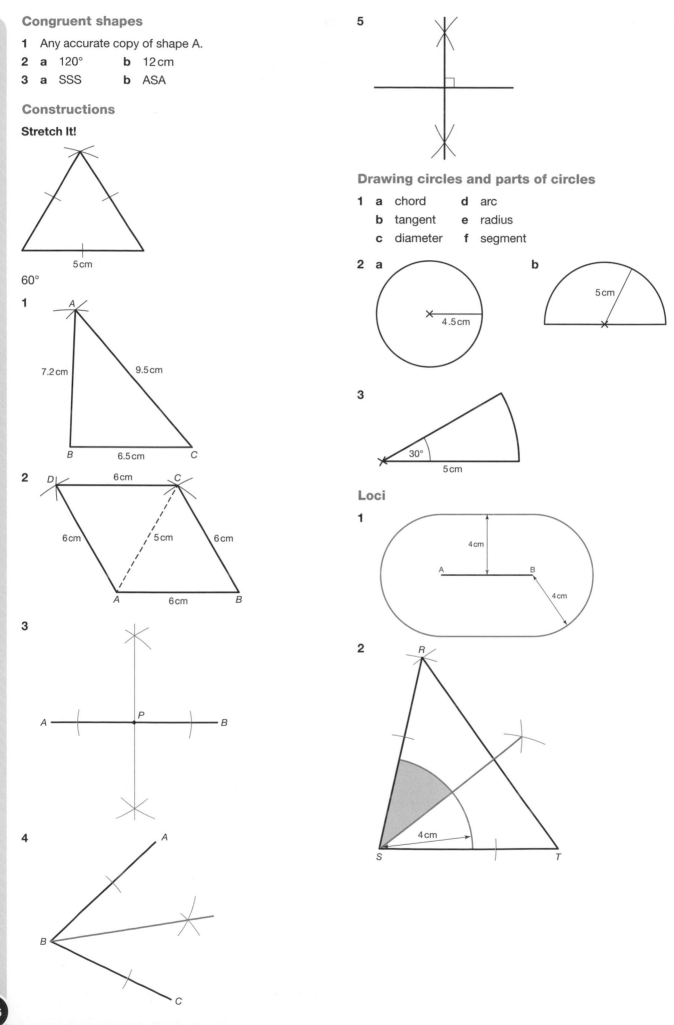

60°

1

2

3

4

5

Drawing circles and parts of circles

1 a chord **d** arc

b tangent **e** radius

c diameter **f** segment

2 a **b**

3

Loci

1

2

3 a and **b**

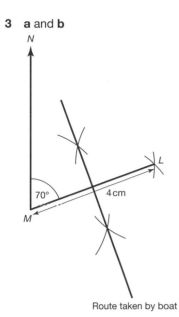

Route taken by boat

Perimeter

1 28.8 cm
2 42 cm
3 $k = 4$, $b = 8$
4 $200 + 30\pi$ m
5 £1.80

Area

Stretch it! Area of a semicircle $= \frac{\pi r^2}{2}$,

Area of quarter circle $= \frac{\pi r^2}{4}$

1 **a** 9.0 cm²   **c** 28.0 cm²   **e** 63.6 cm²
   b 4.5 cm²   **d** 5.0 cm²
2 9 cm²
3 $\frac{1}{4}$
4 18 cm²
5 454.1 cm²

Sectors

1 Area = 39.3 cm²
   Perimeter = 25.7 cm
2 Area = 12π cm²
3 £119.92

3D shapes

Stretch it!

3D shape	Faces	Edges	Vertices
Cube	6	12	8
Cuboid	6	12	8
Square-based pyramid	5	8	5
Tetrahedron	4	6	4
Triangular prism	5	9	6
Hexagonal prism	8	18	12

Stretch it!

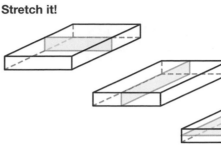

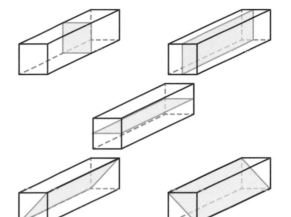

Stretch it!

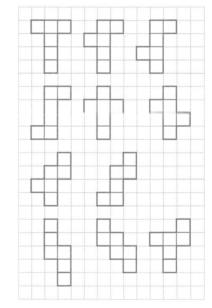

1 **a** 6 possible rectangular faces:

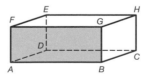
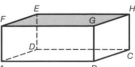
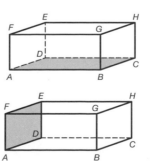
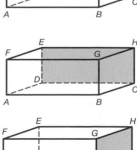

b

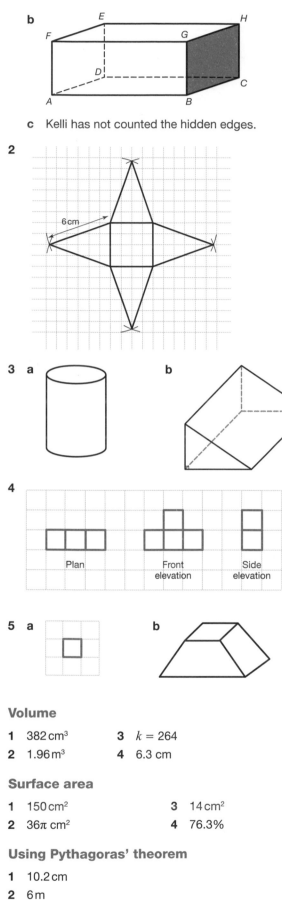

c Kelli has not counted the hidden edges.

4 If the triangle is right-angled, $PQ^2 = PR^2 + RQ^2$

$PQ^2 = 13^2 = 169$

$PR^2 + RQ^2 = 9^2 + 7^2 = 81 + 49 = 130$

$PQ^2 \neq PR^2 + RQ^2$

Claudia is not correct.

5 10.63

6 $\sqrt{13}$

7 £910

8 $DE = 7.2$ cm

Trigonometry

Stretch it! Any lengths for the opposite and hypotenuse in the ratio 1:2.

1 a 0.4	**c** 1.0	**e** 48.6
b 0.6	**d** 26.6	**f** 54.7

2 4.6 cm

3 40.6°

4 $\sin 15° = \frac{\text{opposite}}{10}$

Opposite = 2.6 m

Exact trigonometric values

1 a 0.5   **c** 0   **e** $\sqrt{3}$

  b 0   **d** $\frac{1}{\sqrt{2}}$

2 $AC = 4$ cm

  $BC = 4\sqrt{2}$ cm

3 30° and 60°

4 $\sin 30° = \frac{1}{2}$ therefore, $ABC = 30°$

5 0.5, $\frac{3}{4}$, $\cos 30°$, $\tan 45°$

Transformations

Stretch it! Yes.

1

3 a 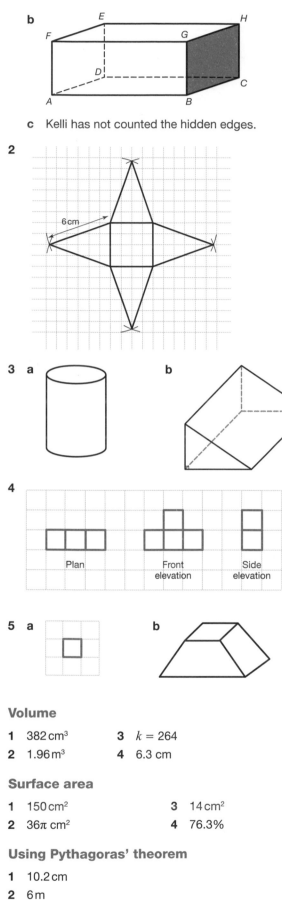 **b**

4

5 a **b**

Volume

1 382 cm³   **3** $k = 264$

2 1.96 m³   **4** 6.3 cm

Surface area

1 150 cm²   **3** 14 cm²

2 36π cm²   **4** 76.3%

Using Pythagoras' theorem

1 10.2 cm

2 6 m

3 54 cm²

2

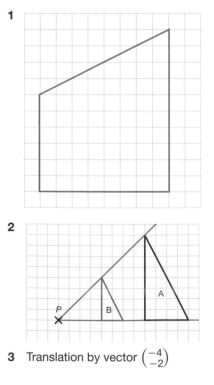

3 Translation by vector $\begin{pmatrix} -4 \\ -2 \end{pmatrix}$

4 a and **b**

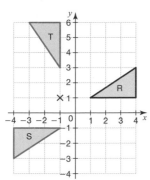

5 Reflection in the y-axis

6 Enlargement by scale factor $\frac{1}{2}$, centre (3, 3)

7 a and **b**

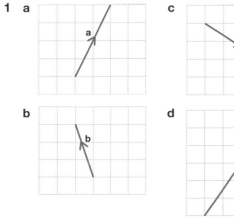

c Rotation of 90° clockwise about (0, 0)

Similar shapes

Stretch it! Perimeter of triangle $ABC = 14$ cm; perimeter of triangle $DEF = 28$ cm. The perimeter of a shape enlarged by scale factor 2 will also be enlarged by scale factor 2. When a shape is enlarged by any scale factor, the perimeter of the shape is enlarged by the same scale factor.

1 a 30° **b** 16 cm **c** 2 cm
2 a 80° **b** 13.2 cm **c** 4 cm

Vectors

1 a 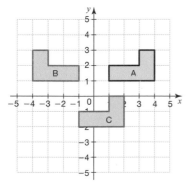 **c**

b **d**

2 a

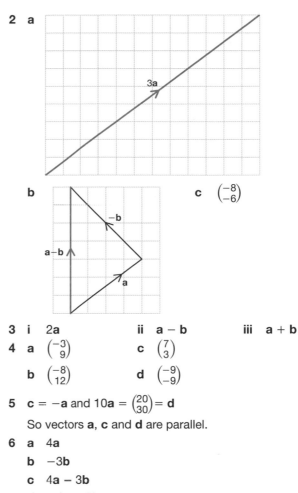

b 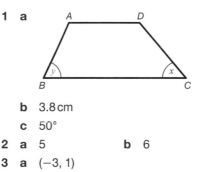 **c** $\begin{pmatrix} -8 \\ -6 \end{pmatrix}$

3 i 2**a** **ii** **a** − **b** **iii** **a** + **b**
4 a $\begin{pmatrix} -3 \\ 9 \end{pmatrix}$ **c** $\begin{pmatrix} 7 \\ 3 \end{pmatrix}$
b $\begin{pmatrix} -8 \\ 12 \end{pmatrix}$ **d** $\begin{pmatrix} -9 \\ -9 \end{pmatrix}$

5 **c** = −**a** and 10**a** = $\begin{pmatrix} 20 \\ 30 \end{pmatrix}$ = **d**
So vectors **a**, **c** and **d** are parallel.

6 a 4**a**
b −3**b**
c 4**a** − 3**b**
d −4**a** − 3**b**

Review it!

1 a

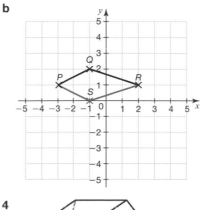

b 3.8 cm
c 50°
2 a 5 **b** 6
3 a (−3, 1)

b

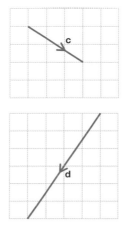

4

5 Any accurate copy of the shape

6 72 cm²

7 24 cm

8

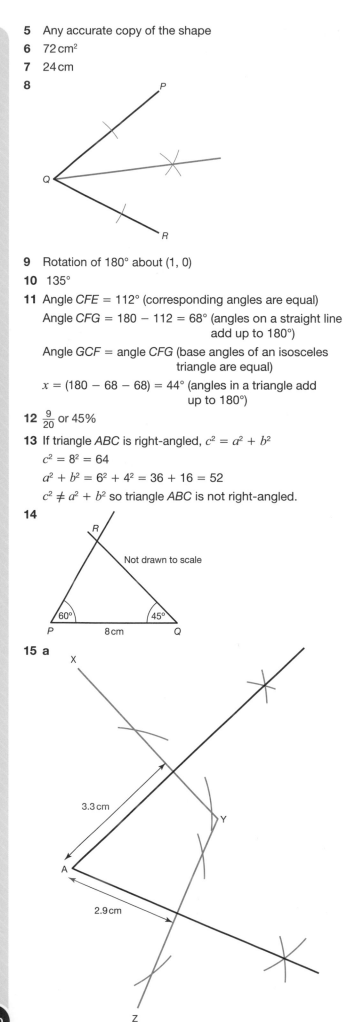

9 Rotation of 180° about (1, 0)

10 135°

11 Angle CFE = 112° (corresponding angles are equal)

Angle CFG = 180 − 112 = 68° (angles on a straight line add up to 180°)

Angle GCF = angle CFG (base angles of an isosceles triangle are equal)

x = (180 − 68 − 68) = 44° (angles in a triangle add up to 180°)

12 $\frac{9}{20}$ or 45%

13 If triangle ABC is right-angled, $c^2 = a^2 + b^2$

$c^2 = 8^2 = 64$

$a^2 + b^2 = 6^2 + 4^2 = 36 + 16 = 52$

$c^2 \neq a^2 + b^2$ so triangle ABC is not right-angled.

14

Not drawn to scale

60°   45°
P   8 cm   Q
R

15 a

b 0.8 m

16 a $\frac{1}{\sqrt{2}}$ **b** $AB = 3$ cm

17 a $\begin{pmatrix} 2 \\ -8 \end{pmatrix}$ **c** $\begin{pmatrix} -6 \\ 13 \end{pmatrix}$

b $\begin{pmatrix} 8 \\ 1 \end{pmatrix}$

18

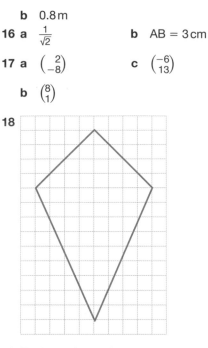

19 Equilateral triangle

20

21 a   i 35°

ii Base angles of an isosceles triangle are equal.

b 110°

c Triangle XYZ is isosceles so angle XYZ = angle YXZ. Angle c = 55°

22 12.6 cm

23 135°

24

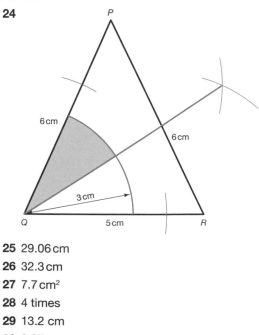

25 29.06 cm

26 32.3 cm

27 7.7 cm²

28 4 times

29 13.2 cm

30 3.87 cm

31 a 300 cm²   **b** 300 cm³

32 53.1°

33 Translation by vector $\begin{pmatrix} -7 \\ -6 \end{pmatrix}$

34 5**a** + 2**b**

Probability

Basic probability

Stretch it! No – each time the probability of getting an even number is $\frac{1}{2}$. You would expect to get even numbers approximately 50 times but cannot guarantee it.

1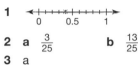

2 a $\frac{3}{25}$ **b** $\frac{13}{25}$

3 a

4 b

5 0.43

6 0.4

Two-way tables and sample space diagrams

1

	Chicken	Beef	Vegetarian
Fruit	12	6	4
Cake	5	3	8
Total	17	9	12

a 12   **b** as above

2 a

		Dice 1					
		1	2	3	4	5	6
Dice 2	1	2	3	4	5	6	7
	2	3	4	5	6	7	8
	3	4	5	6	7	8	9
	4	5	6	7	8	9	10
	5	6	7	8	9	10	11
	6	7	8	9	10	11	12

b i $\frac{1}{18}$ **ii** $\frac{1}{12}$ **iii** 0

3 1, 1, 3, 3

Sets and Venn diagrams

Stretch It! 2

Stretch it! None

1 a
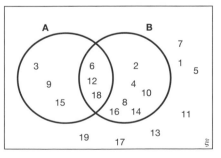

b Multiples of 6

2 a C ∩ T is the set of students who travel by car **and** train.

C' ∩ B is the set of students who do **not** travel by car **and** travel by bus.

b i $P(C) = \frac{19}{50}$

   ii $P(B \cup T) = \frac{3}{5}$ **iii** $P(B' \cap T) = \frac{7}{25}$

3
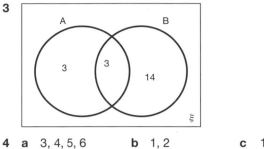

4 a 3, 4, 5, 6   **b** 1, 2   **c** 1, 2, 3

Frequency trees and tree diagrams

1 a
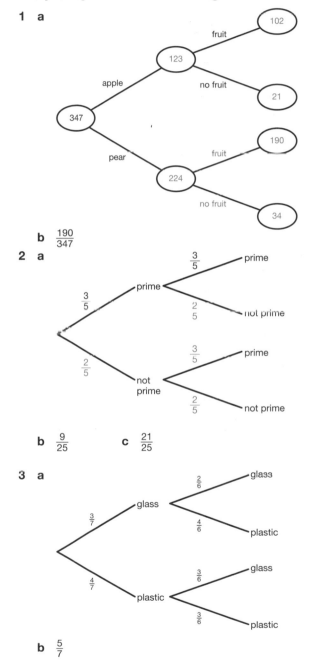

b $\frac{190}{347}$

2 a

b $\frac{9}{25}$ **c** $\frac{21}{25}$

3 a

b $\frac{5}{7}$

Expected outcomes and experimental probability

Stretch it! The dice has not been rolled enough times – carry out further tests.

1 135

2 20 red sweets

3 50 primes

4 a Charlie — he has carried out the most tests.

  b 6

Review it!

1 30

2
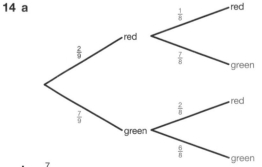

3 0.7

4 B, C

5 a $\frac{3}{5}$ **b** 5

6

	Pizza	Pasta	Risotto	Total
Cake	12	6	1	19
Ice cream	10	11	10	31
Total	22	17	11	50

7 0.7

8 a No, he has not tested his dice enough times.

  b 18

9 0.55

10 a

		\multicolumn{6}{c}{Dice}

		1	**2**	**3**	**4**	**5**	**6**
Coin	Head	2	4	6	8	10	12
	Tail	3	4	5	6	7	8

  b i $\frac{1}{6}$ **ii** $\frac{1}{6}$

11 2, 3

12 a i 6 **ii** 1 **iii** 5

  b $\frac{4}{8} = \frac{1}{2}$

13 a i $\frac{8}{15}$ **ii** $\frac{7}{30}$ **iii** $\frac{7}{10}$

  b No. If you do, you will count P(S ∩ P) twice.

14 a

```
                              1/8   red
                    red ─────<
          2/9                 7/8
         /                          green
        <
         \          7/9
          7/9  green ─────<  2/8   red
                              6/8
                              green
```

  b $\frac{7}{12}$

15 a 10

  b $\frac{1}{3}$ is only a theoretical probability and therefore will not necessarily be accurate in real life.

16 0.74 or $\frac{37}{50}$

17 £50

18 a Milo: he has surveyed a larger sample. **b** 263

Statistics

Data and sampling

Stretch it! A random sample could be taken; you could allocate a number to each student and randomly generate the numbers to survey. Any method is acceptable as long as each person in the school has an equally likely chance of being chosen. Alternatively a stratified sample could be taken.

1 Primary Source; Recording the data by measuring it yourself.

Secondary source: Any sensible source, e.g. the Met Office, local paper etc.

2 Qualitative data.

3 It is cheaper and quicker than surveying the whole population.

4 a The people working for an animal charity are more likely to be opposed to wearing real fur. Every member of the population does not have an equal chance of being chosen.

  b Surveying people in the street, a random telephone survey. Any sensible method that ensure that each member of the population does has an equal chance of being chosen.

5 a $\frac{1}{5}$ **b** 10 bottles

6 a 12 000

  b The sample is relatively small. The sample is not a random sample as it is taken on one day in a year.

7 a Two of the following: the groups overlap; they are unequal in width; there is no group for anyone with a journey of more than 60 minutes; most people will be in the middle group.

  b How long do you spend travelling to school in the morning?

  $0 < t \le 10$

  $10 < t \le 20$,

  $20 < t \le 30$,

  $30 < t \le 40$

  $t > 40$

Frequency tables

1

Number of people on the bus	Frequency
0–9	4
10–19	12
20–29	3
30–39	1

2 a

Number of courgettes	Frequency
0	1
1	0
2	1
3	1
4	9
5	3
6	0

  b 56

3 a There are gaps between the groups.

The groups do not have equal widths.

b Although one of the data values may fall in the $30 \leq h \leq 40$ group, this doesn't mean that they trained for 40 hours. They could have trained for any length of time between 30 and 40 hours.

Bar charts and pictograms

1 a 20   **b** 25%

2 a 4   **b** 39.2%

c Proportion of boys who play two sports $= \frac{6}{18} = \frac{1}{3}$

Proportion of boys who play three sports $= \frac{3}{12} = \frac{1}{4}$

$\frac{1}{3} > \frac{1}{4}$ so the proportion who played two sports is larger.

3

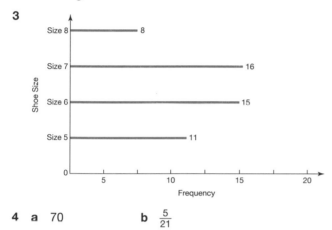

4 a 70   **b** $\frac{5}{21}$

Pie charts

Stretch it! Round appropriately – but check the angles sum to 360°.

1 Spanish = 168°

French = 108°

German = 84°

2 a 1   **b** 97.5% or $\frac{39}{40}$

3 a 28

b The bar chart, since the frequency is easy to read from the bar chart.

Stem and leaf diagrams

1 a 7   **b** 0.2 kg/200 g

2 1 Age of people using a dentist

```
2 | 0 0 0 0 1 1 1
3 | 2 5 5 7
4 | 1 2 2 6
```

The leaves were not in ascending order. The spaces between leaves were not regular.

3 a Stem and leaf diagram – you can see the smallest number of passengers was 3, however on the bar chart you only know it is between 0 and 9.

b Both since the shape of the data is preserved in both.

Measures of central tendency: mode

1 The other three must be 12.2.

2 $1 < t \leq 2$

3 17

4 Max is correct. The modal number of pets is the group with the highest frequency, therefore 2 pets is the mode.

Measures of central tendency: median

1 5.4

2 $2 < b \leq 4$

3 a Group A = 83.5

Group B = 77

b Group A

Measures of central tendency: mean

Stretch it! a Mode   **b** Mean/Median   **c** Mean/Median

1 a Mean = 4.6

b You are using the midpoint of the groups as an estimate of the actual value for each group.

2 8.5

3 No – they could be any pair of numbers which sum to 10.

Range

1 8.8

2 a Girls = 3

b Boys = 2

3 Athlete A

4 35% or 75%

Comparing data using measures of central tendency and range

1 a i Mean = 83.5 minutes

    ii Median = 30.5 minutes

b The extreme values affects the mean but not the median.

2 All the data is used to find the mean.

3 Either as long as suitably justified:

Car A – although the mean time is higher, it is more consistent in performance since the range is smaller.

Car B – the acceleration is quicker on average.

4 a and b The mode or median since the mean will not be a whole number and therefore not meaningful.

Time series graphs

1

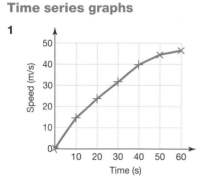

2 a 67°C

b Approx. 27°C

c No, since it is extrapolation (beyond the limits of the data).

3 a 17 000

b i April **ii** August

c The number of tourists peaks in April and again in December. The low seasons are February/March and July/August/September.

Scatter graphs

1 a and **b**

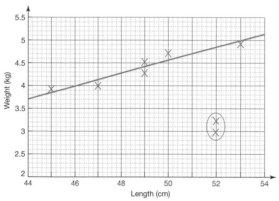

c Positive

d This will vary according to the line of best fit: approximately 4.8 kg

e This is beyond the limits of the data and therefore extrapolation.

2 a, b and **c**

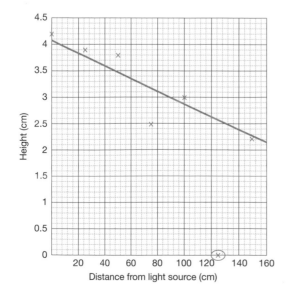

c The seeds failed to germinate or the seedling died.

d The further the seedling is from the light source the shorter its height.

3 No, although the two things correlate one does not cause another. There may be many reasons why the crime rate is high in the area, perhaps there is poverty and inequality causing social tension.

Graphical misrepresentation

1 The organisation has only shown a small section of the data. This is not enough to understand the overall trend.

2 Chart C correctly shows the information. Chart A has an incorrect vertical axis, suggesting there are more women than there actually are. Chart B has unequal bar widths, also exaggerating the number of women.

Review it!

1 The sample is too small and he only asked his friends, therefore not representative of the population of TV viewers.

2 a Margherita **b** 5% **c** 18°

3 a $\frac{1}{4}$

b 480 cars **c** $\left(\frac{105}{360}\right) \times 480 = 140$ cars

4 a The number of people doing their grocery shopping online is increasing.

b Any sensible answer: approximately 75%

c No – it is outside the limits of the data therefore extrapolation.

5 a Outside: **i** Mode = 21 and 31 **ii** Median = 28.5 **iii** Range = 21

Greenhouse: **i** Mode = 47 **ii** Median = 47 **iii** Range = 14

b The seedlings are taller in the greenhouse since both mode and median is larger, the range of data is smaller in the greenhouse so the height the seedlings reach is more consistent.

c Range = 31

6 Comparative bar chart or compound bar chart:

a

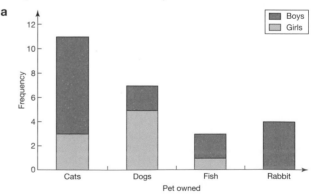

or:

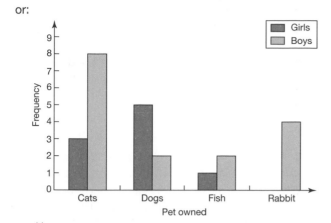

b $\frac{11}{25}$

7 a 40−59

b The youngest person is between 0 and 19, the youngest may be any age in this range and the oldest is between 80 and 99 therefore any age in this range.

8 a 7

b Size 5

c Mean = 5.3̇

d Mode — the mean is not an actual shoe size.

9 a Time for 800 m (seconds)

```
11 │ 2 2 5 8 9
12 │ 0 1 9
13 │ 1 2
```
Key: 11| 2 = 112 seconds

b $\frac{4}{10} = \frac{2}{5}$

10 a $\frac{1}{3}$ **b** 20 **c** Biology

11 a and c

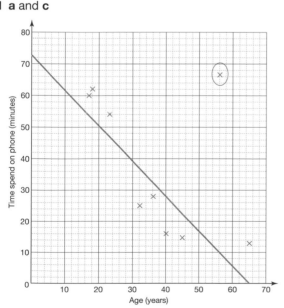

b Negative

d Approximately 40 minutes: depends on line of best fit.

e This is outside the limits of the data and therefore extrapolation.

f As the age of the customer increases the time spent on the phone decreases.

12 a 74

b The midpoint of the class is used as the age of each of the patients rather than the actual age.

13 Male $< 50 = 40$

Female $< 50 = 35$

Male $\geq 50 = 19$

Female $\geq 50 = 26$

14 Annual income for surveyed population

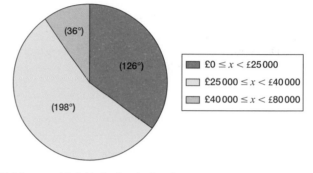

15 Mean = 10.2 (1 decimal place)

16 2, 3, 3, 5, 6